Gender Differences in Technology and Innovation Management

De Gruyter Studies in
Innovation and Entrepreneurship

Series Editor
John Bessant

Volume 2

Gender Differences in Technology and Innovation Management

Insights from Experimental Research

Edited by
Alexander Brem, Peter M. Bican, Christine Wimschneider

DE GRUYTER

ISBN 978-3-11-059079-1
e-ISBN (PDF) 978-3-11-059395-2
e-ISBN (EPUB) 978-3-11-059114-9
ISSN 2570-169X

Library of Congress Control Number: 2019955041

Bibliographic information published by the Deutsche Nationalbibliothek
The Deutsche Nationalbibliothek lists this publication in the Deutsche Nationalbibliografie;
detailed bibliographic data are available on the Internet at http://dnb.dnb.de.

© 2020 Walter de Gruyter GmbH, Berlin/Boston
Typesetting: Integra Software Services Pvt. Ltd.
Printing and Binding: CPI books GmbH, Leck

www.degruyter.com

Preface

The idea for this book surfaced during the International Technology Management Seminar of the Chair of Technology Management at the Friedrich-Alexander-University Erlangen-Nuremberg (FAU), Germany. Every study term, this interdisciplinary and intercultural seminar is focused on scientific methods of data collection and analyses based on different topics of contemporary societal relevance and impact. In the summer term 2018, the seminar concentrated on one of the most pressing topics on contemporary research: The "gendered innovation" gap. Financially supported by the program „Förderung der Chancengleichheit für Frauen in Forschung und Lehre" (promotion of equal opportunities for women in research and teaching) of the Office of Gender and Diversity at the FAU, it enabled setting-up the seminar different to standard seminars: The course was organized as a three-part interdisciplinary project entitled „Technology Management in Practice – Gender Differences in Innovation", involving eleven students from the School of Business, Economics, and Society, as well as the Faculty of Engineering. The three interrelated elements of the project consisted of (1) a scientific part, (2) an empirical-practical part, and (3) a knowledge-transfer part to the general public.

In the scientific part, students reviewed gender relevant literature and identified theories to derive hypotheses on gender differences in various areas of innovation and technology management. Subsequently, a suitable methodology to empirically test the derived hypotheses was developed. The clear grouping of the sample according to the closely matched key variable gender and the prime goal to examine gender-based human behavior and interaction with technologies led to the choice of an experimental design as research method (Gray, 2018). Local technology-oriented firms sponsored technical equipment and innovative products, which were used in the different tasks throughout the experiments. Some firms also had questions of interest concerning gender-specific alignment of their products, which some experiments took upon additionally.

The empirical-practical part formed as data collection event basis. On April 26th, 2018, at the International Girls' and Boys' Day, the experiments were carried out with pupils from a local vocational school, aged between 16 and 20 years. The decision to schedule the data collection on this date was due to the day's special purpose to highlight diversity, equal opportunities, and future employment paths for boys and girls independent of their sex. This message fitted the overall aim of our project extraordinarily well. To switch the standard classroom setting with an innovative and creative work-environment, the experiments took place outside university at a special location: the Zollhof Tech Incubator, a home for tech startups and digital innovators in the metropolitan area of Nuremberg, Germany.

Eventually, the transfer of the experimental results and the derivation of implications for practice and further research was ensured at several levels: First, the students had to prepare scientific papers and present their findings as part of a conference to which representatives from research, politics, and industry were invited. In this way, the findings could be disseminated not only internally, but also create knowledge outside the university's boundaries. The Heizhaus ("heating house") of Nuremberg's former flagship company Quelle, which had been repurposed as co-working space for all kinds of activities, like traditional craftmanship and artisanal business models, served as a venue for this conference. A jury composed of members active in gender-related topics evaluated the student projects' relevance, innovativeness, methodological procedure, and rigor, also accounting for the results' potential of promoting gender equality. Based on these criteria, the best three presentations were awarded. Second, the knowledge transfer into practice took place through a poster presentation of the most intriguing findings in front of the pupils that previously participated as test and control groups of the experiments. The students gave their presentations in a pupil-friendly way to draw attention to particularities in the results concerning male and female participation. This way it was possible to raise awareness for typical stereotypes, which could be disproved and opportunities for both female and male gender, which are not sufficiently exploited yet. Third, and utmost, the ultimate knowledge transfer in order to make the research results available to a broad audience is fostered through this edited book. Hence, this book comprises of a hand-picked selection of research papers resulting from this International Technology Management Seminar. We hope that you, the reader, enjoy the chapters and results as much as we had fun in bringing this project to life, running the seminar, working with our students, and, ultimately, editing and finalizing this book.

Finally, we would like to thank all those who have contributed to make this book possible. A special thank-you goes to Dr. Imke Leicht and Dr. Magda Luthay from FAU's Office of Gender and Diversity, who granted the financial resources from the program "Promotion of Equal Opportunities for Women in Research and Teaching" that enabled this project and awarded the project's results and contribution to the topic of gender equality at the FAU with the Renate-Wittern-Sterzel Price 2018 (the yearly FAU-Award for Gender Equality). The Zollhof Tech Incubator Nuremberg as well as the co-working community Quelle "Heizhaus" hosted the two events that took place during the project. We would like to thank the staff of the two venues who helped to organize and run the events. Many thanks also go to Aya Jaff, Dr. Magda Luthay, Prof. Dr. Brigitte Schels, Corinne Schindlbeck, and Prof. Dr. Silke Tegtmeier for serving as jury members at the conference and evaluating the research projects. We further thank Réka Müller and Theresa Wöllner for their valuable proof-reading and revision of the final chapters. Last but not least, we would like to thank all contributors of this project, as well as the responsible people at

Degryuter, especially to Stefan Giesen. It is also our honor to be part of the De Gruyter Studies in Innovation and Entrepreneurship, edited by John Bessant.

Overall, we hope to stimulate further research on this important topic, and to motivate other researchers to try new formats of teaching with their students.

Alexander Brem, Peter Bican and Christine Wimschneider

Contents

Preface — V

Alexander Brem, Peter Bican, Christine Wimschneider
1 Introduction: Why Gender Differences in Technology and Innovation Matter — 1

Alexandra Fleck, Theresa Wöllner
2 The Impact of Examples on Creativity: Gender Differences in Fixation Effects — 7

Philipp Hässler, Sven Schneider
3 Effects of Gender Differences in Competition on Creativity — 29

Julian Hertzler, Maximilian-Cyrus Mehrpour
4 Gender Differences in Intuitive Product Usage — 43

Sophia Ohmayer, Leonardo Reuther, Bangdi Wang
5 Spatial Reasoning Ability and Methodological Problem-Solving in STEM at an Intersection of Gender — 57

Philipp Röll, Michael Heimes
6 Gender Differences in Approaching and Solving Technical Tasks – An Experimental Research — 73

Stephanie Birkner, Janina Sundermeier, Silke Tegtmeier
7 E-health Value Creation Revisited: Towards a Gender-Aware Typology of Digital Business Models — 87

List of Figures — 105

List of Tables — 107

Index — 109

Alexander Brem, Peter Bican, Christine Wimschneider
1 Introduction: Why Gender Differences in Technology and Innovation Matter

The imbalance in the representation and contribution of women and men to the engineering and technology domain is a phenomenon of our time. Female founders, for example, have more often a social or creative background, but less often come from STEM-related subjects (FFM, 2018). At the same time, it is a glaring contradiction, as, stronger than ever before, technology and innovation are interwoven with everybody's daily life (Delaney & Devereux, 2019). The gender inequality already starts with different excitement and interest for STEM-related subjects at school and continues when choosing a study program or later in the professional sphere of research, teaching or the industry (Bukstein & Gandelman, 2019; Heilbronner, 2013). Women and men are not only subject to differences in the access and selection of education and career pathways but also underlie unconscious social judgements concerning their abilities in the same remit. Boring (2017), for instance, found that performance perception and evaluation corresponds to gender stereotypes independent of actual accomplishments. This can put men at disadvantage concerning typical female tasks, such as organizing and preparing, whereas typical male tasks like technical expertise and leadership skills were rated lower in women.

Research has started to draw attention to the severe consequences and risks that can result from scientific studies when calculating with male variables as the norm while leaving female gender unnoticed. Criado Perez (2019) revealed this data gap in her book "Invisible Women" and points towards multiple cases where consumer goods, medicine, and services are optimally designed for men. For females, however, they too often represent a trade-off in usability and suitability, as women's unique characteristics and preferences were not considered during the development and in market surveys. In the field of medicine and pharmacy, there is only one institution in Germany that offers gender-sensitive research and teaching. The co-founder of this "Gender Medicine", Vera Regitz-Zagrosek from Charité Berlin University of Medicine studies sex-specific preventions, mechanisms, and treatments of diseases. This work of Regitz-Zagrosek and her team is seminal in terms of how to treat men and women individually but equally well (e.g. Regitz-Zagrosek et al., 2006, Regitz-Zagrosek, 2012). Also, in fields more related to innovation and technology, such as in entrepreneurship, there are movements emphasizing gender specifics, for instance, research on gender-related differences in founding intentions (Laspita et al., 2007). Other major findings are that women might experience disadvantages due to overt discrimination, socio-economic positioning and resource access, which in turn can lead to constrained business performance compared to men. This should not to be confused with underperformance. Typical female or male traits and abilities due to differences in socialization do not automatically show to have an influence on business success (Fischer et al., 1993, Marlow & McAdam, 2013). For example, female founders prioritize profitability, whereas male founders prioritize company growth (FFM, 2018).

To the best of the editors' knowledge such explicit insights concerning gender specific differences are still pending for typical research areas in technology and innovation management: What are peculiarities in the creativity processes between men and women? Are their differences in females' and males' acceptance and usage of technological devices differently? Do typical stereotypes concerning technical mastery of men vs. females' inferiority still manifest in actual R&D settings? Against the background of these questions, the idea arose to bring the issue of gender to innovation and technology management. With this book we would like to contribute to this ongoing debate. The chapters presented in this book aim to shed light on various aspects of this debate. In particular, this book will cover gender-specific findings concerning creativity and creative problem-solving, competition and collaboration, technical problem-solving, product perception, usage as well as the gender-awareness in business models.

The contents of each book chapter are summarized in the following.

Focusing on one of the core activities in STEM-related work environments, Fleck and Wöllner are concerned with gender differences in creative problem-solving in the **second chapter** of this book. Particularly, they analyse the impact of the fixation effect during creative problem-solving tasks. The fixation effect describes "…situations in which previous experiences or the presentation of examples can have a negative, counterproductive effect on the outcome of cognitive processing because the previously generated ideas interfere with the ability to generate future ideas" (Fleck and Wöllner, Chapter 2). Prior research identified variations of the fixation effect as to age, academic background, or individual experience, hence missed to account for differences in fixation effect outcomes between women and men. During their empirical study with 80 participants, the impact of visual and verbal impulses during the idea generation phase on creativity was analyzed. Amongst others, the results of this chapter indicate that gender differences with regard to the fixation effect exist: Exposure to potential solutions or examples to a task influences women less than men.

Hässler and Schneider investigate gender differences in competitive and creative behaviour in **Chapter 3**. They show that these two factors, competition and creativity, have been explored individually before, yet, research has not investigated gender differences in both fields more thoroughly. However, prior research suggests that women prefer non-competitive settings, while the opposite holds true for men. The results of this chapter on creative behavior under competitive circumstances revealed that men and women produce a fairly equal amount of ideas in non-pressure situations, meaning that creativity is largely independent of gender. Men, however, increase their performance significantly in competitive, mixed-gender group situations, while the performance of women only slightly increases. As work-life is characterized by a high degree of competition, men might feel more comfortable in the current environment, hence women might be penalized. Consequently, Hässler and Schneider propose a change in thinking, which could contribute to a better integration of women in the work place. This chapter shows that competing in teams or single-gender competitions can already contribute to decreasing this gap. As a consequence of their findings, Hässler and Schneider question the role of fixed quotas for women, since long-lasting effects might be achieved more efficient through the proposed change in thinking on competitive behaviour and the consequences for women in the workplace.

In **Chapter 4**, Hertzler and Mehrpour analyse whether gender influences the intuitive use of products. The quasi-experiment that was used to test this question confronted participants with the task to find as many functions of an office chair as possible without any prior instructions or explanations. Male participants were generally able to identify more functions of the office chair than female participants. However, it was also perceived that males had more experience in handling office chairs before. Consequently, they felt more familiar when handling the presented type of office chair. Thus, the results indicate that the intuitive use of a product relies less on gender of the participants, but moreover on other factors, such as prior experience or technological knowledge. As confirmed through prior literature, factors like education, culture, and social environment influence the behaviour and characteristics of men and women. In line with the findings from this quasi-experiment, this gender gap seems not of naturally given origin, but to stem more from past experience and the way in which men and women have been raised and educated.

Ohmayer, Reuther and Wang in **chapter 5**, look at spatial reasoning and methodological problem-solving abilities of men and women to answer the question of why women are still majorly underrepresented in the STEM disciplines. Thereby, spatial reasoning is defined as "…the ability to comprehend, recall or even generate two- and three-dimensional figures and mentally rotate and transform these objects," whereas methodological problem-solving abilities "…could be described as compliance with relevant norms that are important for the respective field." This said, Ohmayer, Reuther and Wang demonstrate in their chapter that men have more pronounced spatial reasoning abilities, while women are better at solving a problem methodically and efficiently. In their experiment, participants were tasked with building a buoyant device that could hold a dummy object above water with the help of forming clay and different assisting materials. The results of the experiment indicate that a reason for such few women currently working in the STEM fields could be in the profound needs of spatial reasoning abilities to successfully engage in STEM-related work-environments. They recommend, due to the significance of this skill, to foster and develop spatial reasoning skills in males and females alike. Especially and in contrast to methodological problem-solving, spatial abilities can be improved by using several tools and methodologies, like virtual and augmented reality or three-dimensional video games, as identified by prior literature (Feng, Spence, Pratt, 2007).

Consequently, private and public education should stimulate the growth of spatial reasoning abilities with the aim to 'reduce the leakage' along the STEM-pipeline.

The **sixth chapter** by Röll and Heimes discusses another STEM-relevant topic in how the approach to and the solving of technical tasks differs between men and women. For this purpose, an experiment was designed in which pupils of both gender had to assemble an office chair, as well as crafting a corresponding instruction for the process in written form. Male participants assembled the chair faster and more effective while female participants wrote more complete and better structured instruction manuals. Besides reasons also presented in the previous chapters of this book, such as spatial thinking, other factors like physical differences between men and women need to be accounted for as well to understand why women are underrepresented in STEM-related fields. For example, to successfully assemble the chair during the experiment conducted in this chapter, a minimum input of physical power is necessary, which some of the female participants were not able to provide physically. These results indicate that collaboration between men and women or mixed team-compositions in STEM-related projects is essential for a comprehensive and successful problem-solving.

As the previous chapters focused on experiments and on identifying effects of gender differences on STEM-related outcomes like creativity or problem-solving skills, the last **chapter 7** of the book explores the issue of how gendered innovation potential can be exploited for the benefit of value creation. In this chapter, Birkner, Sundermeier and Tegtmeier conceptualize a heuristic scheme to analyse the gender-awareness of business models, with a special emphasis on e-health solutions in the context of entrepreneurship. They show that innovations in general, like e-health, and gendered innovations are rarely linked, disregarding the potential of synergetic co-creation of gender-aware value propositions. With their chapter, they aim to pave the way for other approaches that analyse business models from a gendered perspective. In doing so, they (1) extend the existing evidence concerning the relevance of gendered innovation for research endeavors to the area of entrepreneurship research, (2) extend the knowledge on business models to research that lies at the intersection of digital transformation and innovation needs and potentials in (public) health, and (3) conceptualize a heuristic scheme that paves the way for further research into how gender-awareness is manifested in digital business models. With these contributions at hand, practitioners, and here especially entrepreneurs, might want to make use of the classification scheme for a deeper understanding of the content and structural choices that have to be made about gender-aware value creation and, ultimately, gender-aware business models. Furthermore, investors and policy makers alike might consider the heuristic scheme in informing gender- and equality-related decisions.

What does this ultimately imply for research and practitioners alike? Summarizing the main findings and insights created from the studies above, it becomes clear that men and women behave, react and interact differently in typical situations and tasks related to innovation and technology management. Some contexts seem to impact women less than men (e.g. influences and fixation during idea-generation) while in others, such as in competitive environments, there is evidence that men benefit compared to women. Being aware of these preferences and understanding their dynamics and consequences might help a lot to introduce more effective policies and workplace designs. Thus, research is advised to tap into these very specific areas and derive innovative ideas how the gender gap in pay, management positions and representation in certain fields can be closed aside from quotas. Besides these contrasts in males and females, there are also factors that mitigate potential advantages and disadvantages between genders. Prior experience in the matter and building of knowledge lead to better performance in solving technical problems. Therefore, continuous confrontation with and education in STEM related subjects already starting at an early stage cannot be considered highly enough to bring men and women on the same level. In two studies it became clear that men and women express strengths in different but complementary abilities for solving technical problems. The tendency that the two genders complement each other in their particular approaches, way of thinking and techniques finally provides a positive fundamental idea of the gender debate: Embracing the differences and combining the best of the two spheres. There is still a long road to walk down, however, first promising signs become already visible in female entrepreneurial ventures: Even though female founders are still underrepresented in technical sectors and STEM subjects, a change of prospects and attitude might be on its way: Not only are women more and more represented as (co-)founders in start-ups (28% of all start-ups in 2018), more women are also founding their own start-up than before (FFM, 2018). This makes us believe that contemporary R&D and technology management principles such as interdisciplinarity, openness, and agility pave a suitable way for empowerment and closing of the gender divide, especially in STEM-related areas.

References

Boring, A. (2017). Gender biases in student evaluations of teaching. *Journal of Public Economics*, 145, pp. 27–41.
Bukstein, D. & Gandelman, N. (2019). Glass ceilings in research: Evidence from a national program in Uruguay. *Research Policy*, 48 (6), pp. 1550–1563.
Criado Perez, C. (2019). Invisible Women: Exposing Data Bias in a World Designed for Men, Chatto & Windus, London UK.
Delaney, J.M. & Devereux, P.J. (2019). Understanding gender differences in STEM: Evidence from college applications. *Economics of Education Review*, 72, pp. 219–238.
FFM. (2018). *Female Founders Monitor 2018*. Bundesverband Deutsche Startups e.V. Accessed July 15, 2019 at: https://deutscherstartupmonitor.de/fileadmin/ffm/ffm_2018/Studie%20Female%20Founders%20Monitor%202018.pdf.
Fischer, E., Reuber, R. & Dyke, L. (1993). A theoretical overview and extension of research on sex, gender, and entrepreneurship. *Journal of Business Venturing*, 8(2), 151–168.

Heilbronner, N. N. (2013). The STEM Pathway for Women: What Has Changed? *Gifted Child Quarterly*, 57(1), 39–55.

Laspita, S., Scheiner, C. W., Chlosta, S., Brem, A., Voigt, K. I., & Klandt, H. (2007). Students' attitude towards entrepreneurship: Does gender matter. Rev manag comp int, 8(4), 92–118.

Marlow, S. & McAdam, M. (2013). Gender and Entrepreneurship: Advancing Debate and Challenging Myths; Exploring the Mystery of the Under-Performing Female Entrepreneur, *International Journal of Entrepreneurial Behaviour & Research*, 19 (1), 114–124.

Price, C.A., Kares, F., Segovia, G. & Loyd, A.B. (2019). Staff matter: Gender differences in science, technology, engineering or math (STEM) career interest development in adolescent youth. *Applied Developmental Science*, 23 (3), pp. 239–254.

Regitz-Zagrosek, V. (2012). Sex and gender differences in health. *Science & Society Series on Sex and Science*, 13 (7), 596–603.

Regitz-Zagrosek, V., Lehmkuhl, E. & Weickert, M. (2006). Gender differences in the metabolic syndrome and their role for cardiovascular disease. *Clinical Research in Cardiology*, 95 (3), 136–147.

Alexandra Fleck, Theresa Wöllner
2 The Impact of Examples on Creativity: Gender Differences in Fixation Effects

1 Introduction

1.1 Problem definition and objectives

In the global business environment, the development of new products contributes significantly to the long-term success of a company and ensures its survival (Anderson, Dreu, & Nijstad, 2004; Anderson, Potočnik, & Zhou, 2014). Novel value propositions are particularly beneficial for companies if they clearly distinguish themselves from existing ones and constitute discontinuous innovations. Considering the technological dimension of innovation, such discontinuities are often achieved when firms are moving away from existing technological knowledge and established product technologies (Utterback, 1994). Thus, the degree of innovation of new products is linked to the ability of a company to develop products based on new technological paradigms and solution principles. The foundation for the development of such innovative products must already be laid during the idea generation phase (Cooper, 1988; Henard & Szymanski, 2001), because all subsequent steps will be based on the results of this period (Goel & Pirolli, 1992; Guindon, 1990; Purcell & Gero, 1996).

In this phase of idea generation, creative processes play a particularly important role. The development of new products requires creative work from engineers, designers, and technologists. Therefore, creativity has been considered one of the key qualifications to meet the demands of a modern, constantly changing society (Thompson, 2003). Innovations in science, business and culture are hardly imaginable without creative people. While the potential to be creative exists within each person, there is substantial individual variability concerning the type and the amount of creative output that is generated. A significant subset of empirical studies on creativity focuses on identifying which variables have an impact on creativity and the creative design process.

Examples that are provided during problem-solving, for instance, can influence creative idea production on the individual level. The people that provide these examples may consider them to be clues that assist problem solvers in achieving the desired objectives by suggesting the transfer of relations, heuristics or algorithms from the examples to the particular task at hand (Smith, Ward, & Schumacher, 1993). Examples are often used in professional environments that require problem-solving, such as architecture or engineering. In open situations, where new creations are required to fulfill a desire, examples may be given to illustrate useful approaches of the past or the pitfalls discovered in previous related work. It is obvious that it is generally not the intention of examples to hamper the creative process, but rather to stimulate the generation of beneficial ideas.

However, there are situations in which previous experiences or the presentation of examples can have a negative, counterproductive effect on the outcome of cognitive processing because the previously generated ideas interfere with the ability to generate future ideas. Thus, if knowledge is applied improperly and ideas are unnecessarily restricted, the creative performance might be harmed or even fail (Smith et al., 1993).

Such transfers of experiences and examples have been referred to as the fixation effect. According to Jansson and Smith (1991, p. 3), this can be defined as "a blind adherence to a set of ideas or concepts limiting the output of conceptual design". People become fixated on existing examples to the extent that they are not able to imagine any other way to solve a specific issue. The inhibition of generating new solutions caused by a block or fixation on previous ideas is a common theme in the problem-solving literature (e.g. Duncker & Lees, 1945; Jansson & Smith, 1991; Smith & Blankenship, 1991). Studies have already proven that the fixation effect varies with age, academic background or experience, depending on the individual (Agogué, Poirel, Pineau, Houdé, & Cassotti, 2014; Bonnardel & Marmèche, 2004). However, differences in the fixation effect between men and women have yet to be investigated. Within the literature of creativity, gender differences are inconsistent and generally rare (Baer & Kaufman, 2008). Although substantial proof of differences in patterns and areas of strengths between the sexes exists, there remains relative equality in creative ability. The question arises whether the introduction of examples of solutions during the creative process influences men and women in a similar manner. This issue reveals a gap in the interdisciplinary research on creativity and the fixation effect which the present work intends to close. The overall objective of this study is to gain further insights into the fixation effect and its impact on males and females. For this purpose, this chapter seeks to answer the following research question: Are there differences between women and men when exposed to potential solutions to a creative task that lead to the fixation effect?

1.2 Course of investigation

In order to answer the research question, this chapter is structured as follows. Section 2.2 presents theoretical foundations and preliminary considerations that have decisively influenced the present research work. First, the general concept of creativity is defined and the differences – or rather no differences – between men and women in this area are addressed. Thereupon, the construct of the fixation effect is examined in more detail. This theoretical framework is then used to develop testable hypotheses that will be used to investigate the research question.

The next part of the section consists of the empirical study in which an experiment was conducted with children and adolescents of both sexes to investigate the impact of visual and verbal examples on men and women when confronted with those impulses during the idea generation phase. Section 2.3 describes the methodology of this experiment and provides information on data gathering and research procedures. Key descriptive observations derived from the experiment are then presented in the next section.

In the last section, the key findings are summarized from a scientific perspective. In order to ensure the interpretation of the obtained results, critical points and unavoidable limitations are disclosed from which approaches for further research efforts are derived subsequently. The chapter concludes by deriving implications for practice.

2 Theoretical background

2.1 Creativity

Creativity, like innovation, is inherently interdisciplinary, because it requires the synthesis of knowledge from diverse areas (Amabile & Khaire, 2008; Rhodes, 1961). Although creativity has long been a 'buzzword', there is still no commonly agreed upon and harmonized definition. As with motivation, intelligence and time, the term creativity has its corresponding connotation in everyday language and scientific communication. In general, however, only a small part of the complex scientific phenomenon of creativity has found its way into everyday language, namely the emotional component (Rhodes, 1961). In everyday life, being creative is often equated to being emotionally free, relaxed, uninhibited or free of censorship.

Furthermore, a generally applicable definition of the concept of creativity is only slowly developing in the scientific community. According to Steiner (2011), creativity is the specific ability of an individual or a collaborative system, such as a group, to provide original performance. Likewise, Csikszentmihalyi (1999) defines the concept of creativity as any action, idea or thing that changes an existing domain or transforms it into a new domain. Over time, a multitude of definitions for creativity, some of them highly divergent, have developed, each one emphasizing different aspects of this construct. Thus, the impression could arise that creativity is intangible but quite far-reaching and complex (Sternberg & Lubart, 1999; Taylor, 1988).

However, in order to address the research question, it is necessary to classify and describe the phenomenon as concretely as possible. Therefore, the chapter refers to the definition established by Amabile (1983) which has already been applied in scientific work with a similar focus: Creativity has been defined as the generation of ideas, insights, or solutions that are both novel (to an individual, a group or the world) and useful in solving the problem at hand. Therefore, creative design implies the active change or rejection of previously accepted ideas that might otherwise block progress (Simon, 1996). A creative response or product is one that is determined to be both original and relevant (Runco & Jaeger, 2012; Stein, 1953). The degree of originality is defined by its novelty, uniqueness or statistical rarity, while its relevance is assessed in terms of functionality, usefulness or suitability for a particular objective or in a particular context.

The issue of gender differences in creativity is a topic of considerable scientific and public interest, but also a source of substantial controversy. Due to the heterogeneity of the findings associated with this field of research, the overall picture often appears puzzling or obscure (Baer & Kaufman, 2008). The role of gender in creativity has been studied to determine not only whether men and women differ in their creative abilities or output, but also what factors influence the likely differences and whether these develop in a distinct way over the course of a person's life, as recent studies have shown (e.g. Bender, Nibbelink, Towner-Thyrum, & Vredenburg, 2013; Cheung & Lau, 2010; He & Wong, 2011; Hong & Milgram, 2010; Karwowski, Lebuda, Wisniewska, & Gralewski, 2013; Kaufman, Baer, Agars, & Loomis, 2010; Sayed & Mohamed, 2013; Stoltzfus, Nibbelink, Vredenburg, & Hyrum, 2011). Lack of differences between males and females is the most common result of those studies. However, it is important to emphasize that if there was an overall 'winner' in these studies, women and girls would surpass men and boys (Baer & Kaufman, 2008). Therefore, the authors of the present chapter hypothesize:

Hypothesis 1: Men are more affected by the fixation effect than women.

2.2 Fixation effect

The development of novel, innovative products requires creative thinking skills. Design processes, whether systematic or intuitive, are expected to unleash this creativity by preventing pre-determination or fixation on a specific representation of the design problem or on possible solutions to that issue. Nevertheless, it can happen that people become "'set', 'blinkered' or 'blinded'" when developing new ideas (Crilly, 2015, p. 54). During the cognitive process, people often experience mental fixations that can inhibit individual creativity. The term 'design fixation' is often used to refer to this broad set of phenomena. According to Jansson and Smith (1991, p. 3), it can be defined as "a blind adherence to a set of ideas or concepts limiting the output of conceptual design". Schulthess (2012) differentiates between a functional fixation and a fixation based on existing knowledge, also known as problem-solving set or design fixation.

Functional fixation constrains the cognitive analogy process according to the original purpose of the object of observation. This functional fixation is explained by

Duncker and Lees (1945) through an experiment: The test required the subject to attach a burning candle to a corkboard hanging on the wall while making sure that the wax would not drip onto the floor. The test subjects were only allowed to use a book of matches and a box of thumbtacks along with the candle (Duncker & Lees, 1945). The solution was to separate the box from its contents and place the candle in the box while using the thumbtacks to fix the box to the cork board. During the experiment, most subjects were unable to detach themselves from this functional fixation of the box as packaging material, which is why they failed the task. Other studies revealed corresponding results based on these findings (Adamson, 1952; Adamson & Taylor, 1954).

The problem-solving set refers to the mental fixation when a given problem is solved by re-applying a known procedure. In the study by Luchins (1942), this fixation effect was demonstrated as subjects repeatedly solved complex tasks successfully using a known procedure. Subsequently, the test persons were given a much simpler task, which they solved by applying the already known, but unnecessarily complex solution procedure. Thus, by focusing on previous successes, the participants were not able to recognize an easier and more efficient solution (Luchins & Luchins, 1969).

Jansson and Smith (1991) examined a similar form of fixation, namely the concept of design fixation, in their experiments. The test subjects were assigned the task of developing a bicycle carrier for a specific type of car. Prior to the elaboration phase, the participants were given concrete input on the topic by being presented with solutions that already existed. As expected, the solutions that were then developed mostly resembled the presented approaches. This showed once again that the application of knowledge without further reflection may lead to a counterproductive fixation on already existing solutions. Similar results were obtained by Condoor and LaVoie (2007), Marsh, Landau and Hicks (1996), Marsh, Ward and Landau (1999) and Smith et al. (1993), who also emphasized the influence of individual cognitive limits and opportunities.

The majority of these studies of inspiration and fixation have in common that they are primarily concerned with the idea generation phase of the design process. They typically manipulate certain variables that are related to the stimuli that are presented to the test subjects, such as novelty or the number of stimuli, in order to examine how different characteristics of external stimuli affect the design process and the resulting outcomes (Vasconcelos & Crilly, 2016). Research has demonstrated, inter alia, that textual stimuli can help participants to increase their originality (Goldschmidt & Sever, 2011). It was also found that designers are more inclined to work visually (Hanington, 2003). Thus, they tend to be more inspired by visual stimuli (Goldschmidt & Smolkov, 2006) but are simultaneously more prone to the negative effects from them (Chrysikou & Weisberg, 2005). In addition, Sarkar and Chakrabarti (2008) substantiate the claim that non-verbal representations, like images or videos, are more effective triggers for inspiring solutions than verbal examples. Thus, the authors of the present study predict:

Hypothesis 2: Visual examples influence both men and women more than verbal examples.

The literature on the subject of fixation is continuously expanding and comparisons of the impact on various groups have already been made. For instance, it was examined if the impact of the fixation effect varies in its intensity at different training levels (undergraduate students, professionals) (Jansson & Smith, 1991). Furthermore, it has previously been investigated how the presentation of examples affects people of academic backgrounds (engineering students, entrepreneurs, designers) (Agogué & Cassotti, 2013). The authors utilized the classical creative ideation task 'the egg task' for this purpose. It consists of proposing a maximum number of solutions within ten minutes to ensure that a hen's egg dropped from a height of ten meters does not break. The subjects were randomly assigned to one of two experimental conditions: One control condition without an example and one test condition with an example of one possible solution. The problems were identical in all conditions except that the group with an example read the following: "One solution classically given is to slow the fall with a parachute". The results showed how the nature of the fixation effect evolves with age. Adolescents were more able than children to explore alternatives outside of the fixation effect using expansive reasoning. Thus, the authors of the present work hypothesize:

Hypothesis 3: Adolescents, no matter what gender, are less influenceable than children.

3 Methodology

To test these three hypotheses that were established on the basis of theoretical considerations, a set of experiments was conducted in which the impact of external appeals on creativity was tested. The experiments were carried out with 80 test subjects and were divided into two different testing days according to age. The first group consisted of adolescents between 16 and 20 years and the second group included children between the ages of six and eleven. The first experiment took place at the Girl's and Boy's Day and the participants were drawn from a local high school. Their background was either social or economic. The second experiment took place at the vacation day-care, the "FAU Pfingstferienbetreuung", at the Friedrich-Alexander-University Erlangen and all the children are related to University members such as doctors, engineers and scientific personnel. The mean of the ages according to the groups is shown in Table 2.1.

Table 2.2 shows the distribution of males and females according to age. Due to illness, the attendance rate is not evenly divided between the genders and ages. It

Table 2.1: Overview of the test subjects (Source: Own illustration based on Agogué and Cassotti, 2013, p. 6).

Group	N	Age (mean)	Education
Children	48	7.96	Elementary school
Adolescents	32	17.25	High school

Table 2.2: Distribution of the genders (Source: own illustration based on Agogué and Cassotti, 2013, p. 6).

Group	Female	Male
Children	23	25
Adolescents	16	16

is recommended to have an equal amount of probands for each category to ensure a better comparison.

The experiments happened consecutively and the task was conducted silently and individually. The participants were instructed to keep their eyes on their own work. Everybody had the same devices such as a modeling stick, a toothpick and four packages of modeling clay in different colors. The clay is called Fimo from Staedler, a German manufacturer of office supplies. During processing, Fimo is soft like modeling clay is supposed to be. Afterward it can be baked so that the material cures. This process is irreversible and the clay can be used as a figure or decoration. The modeling clay's packaging consists either of plain plastic wrap to avoid air contact or a themed box from the manufacturer, called Fimo kids form&play sets. These specific sets are simply called boxes in the following. The topics of the boxes are gender-related as well as neutral. For the girls, the boxes include butterflies, flowers and mermaids whereas the boys' boxes contain dinosaurs and pirates. The neutral one is concerned with the topic farm animals. The colors were partly gendered such as glitter white, pink, rose or black, brown, and white as well as non-gendered colors such as purple, green, orange, blue and red. In every experiment, most of the test subjects either got the box that fit to their gender or the neutral one. However, in every group, one to two subjects were handed the opposite box, e.g. a flower box for a boy or a pirate box for a girl. The groups had 30 minutes to craft something and answer a previously announced questionnaire afterward. The reading out of the instructions for the task was not included in this time period. 30 minutes had already proven to be a perfect amount of time for this task while the experiment was tested with other members of the seminar (not included in the results). The participants were neither too stressed from time-pressure, nor did they become bored with crafting. The probands could use all the clay but they did not have to. For a better understanding, the order of the experiments is depicted in

Figure 2.1. The experiment is structured into four small test sections according to the given hint. The splitting is described on the second and third level of Figure 2.1. The experiments are more thoroughly described in the Methodology.

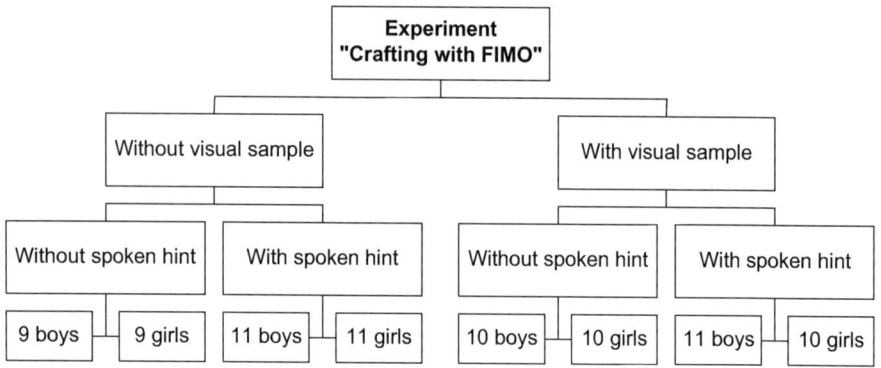

Figure 2.1: Overview of the experiments (Source: Own illustration).

After every experiment, the participants were given a questionnaire with several questions about their background and what they had crafted, e.g. what they made and why, whether they had worked with Fimo or any other kind of modeling clay before, etc. For documentation purposes, the crafted items have been photographed.

First Experiment
The first group is a control group and the test subjects were given the modeling clay without any advice or hint. The modeling clay has the same colors as the clay in the boxes. Every group was given the same instructions, which were outlined before the experiment:

> We do not care about what you make, you can make whatever you want, but it is important that in the end something is created. You are allowed to take your handicraft home. At the place you can see several tools which might help with fine-tuning. You will have enough time, do not feel pressured. You can use all the clay, but you do not have to. We will not answer questions during the experiment.

The intention behind allowing the test subjects to take their work home with them is to motivate the pupils to craft something that they really would like to have. The control group is used to draw attention to hypes or trends in the market that could affect the creativity of the test subjects.

Second Experiment
The second group had the same conditions as the first one. The only difference was that they were given a spoken hint. This hint must refer to an attraction or an event

which is relevant and present to everybody, independent from their gender or age. Thus, Mother's Day is an appropriate hint and, in addition, most of the test subjects associate a Mother's Day gift with something self-made. The favorite movie or series was chosen in the second experimental setup with the younger children because Mother's Day had already passed. This hint fulfills the same function as the first one. Before that experiment, the test subjects were made to listen carefully to the spoken instructions to ensure that everyone understands the information.

Third Experiment
In the third experiment, the test subjects received the modeling clay, together with the packaging box. The instructions were the same, but without the spoken hint. The box included clay, the instructions on how to build the figures on the pictures, the modeling device and a picture related to the theme of the box, that can be used as a background for the figures or work. The pictures printed on the top of the theme boxes are visual samples. The instruction is a step-by-step description on how much clay to use and how to form it. In addition, the box contains advertisements for different boxes with pictures as well.

Fourth Experiment
The fourth experiment was a combination of the second and the third setup. In addition to the instructions, the test subjects were confronted with the visual and spoken example. The same boxes are chosen as well as the same spoken samples. To make sure that everybody understands the spoken hint, the boxes were distributed after the instructions were given to ensure that everybody pays attention.

4 Results

The results of a questionnaire conducted before the actual experiment revealed that all test subjects had crafted with clay before, so the abilities and previous knowledge are comparable. Only 30% had experience with Fimo, but this does not affect the results of the experiment. The assessment of the crafting and whether the person showed a fixation on the provided hint is based on two factors. On the one hand, the appearance of the crafted item was evaluated and on the other hand the answers of the questionnaire. If any of the factors applied, the craft was considered to be influenced by the hint. Appearance was especially decisive for the evaluation of the behaviour of the younger children, because most of them were not able to reflect their actions properly. For example, many crafted a dinosaur but could only give their personal interests or nothing as a reason.

The first group served as a control group to isolate possible trends in the market that might influence the experiment. The results from this group indicate that there are no relevant trends among the pupils, because all the crafted items were very different. If a few of the test subjects had formed the same or similar item from the modeling clay, independently from each other, that could have pointed towards a trend that is especially prevalent in the participants' minds.

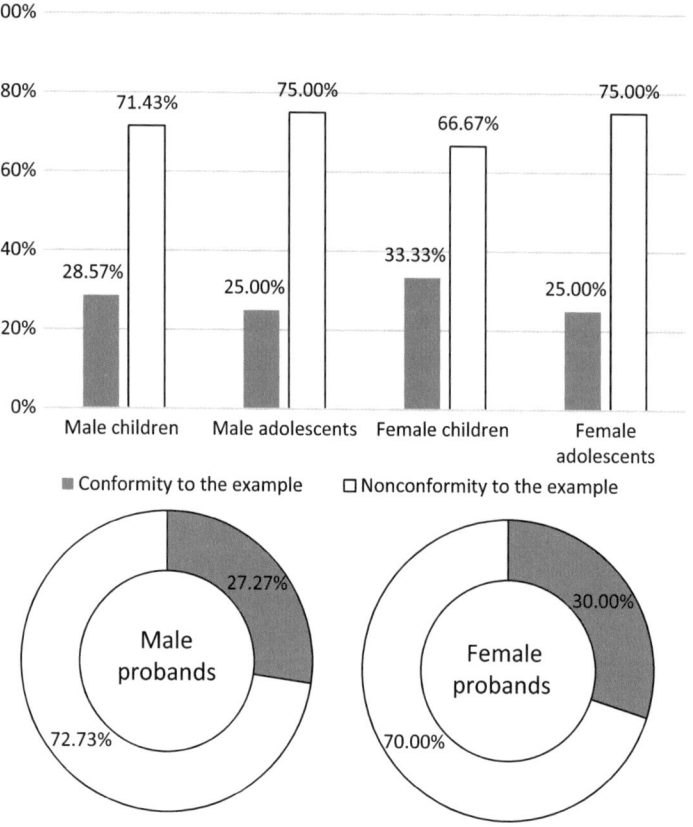

Figure 2.2: Results of the second group (Source: Own illustration).

The results of the second group show that the spoken hint had an influence on the test subjects. The conformity to the spoken hint is shown in percentage, as well as how much it influenced the outcome. Not only are the genders being compared, but also their ages. The circle graphs show the results divided only by gender without considering their age. About 30% of all subjects were influenced by the spoken sample. An example for the Mother's Day hint is a heart or a flower and for the favourite movie it was e.g. the princess from Disney's Frozen. There are no significant differences between boys and girls or between the different ages in this category.

Figure 2.3 shows the results of the third experiment which only included the visual example printed on the box. There is a difference between the two different

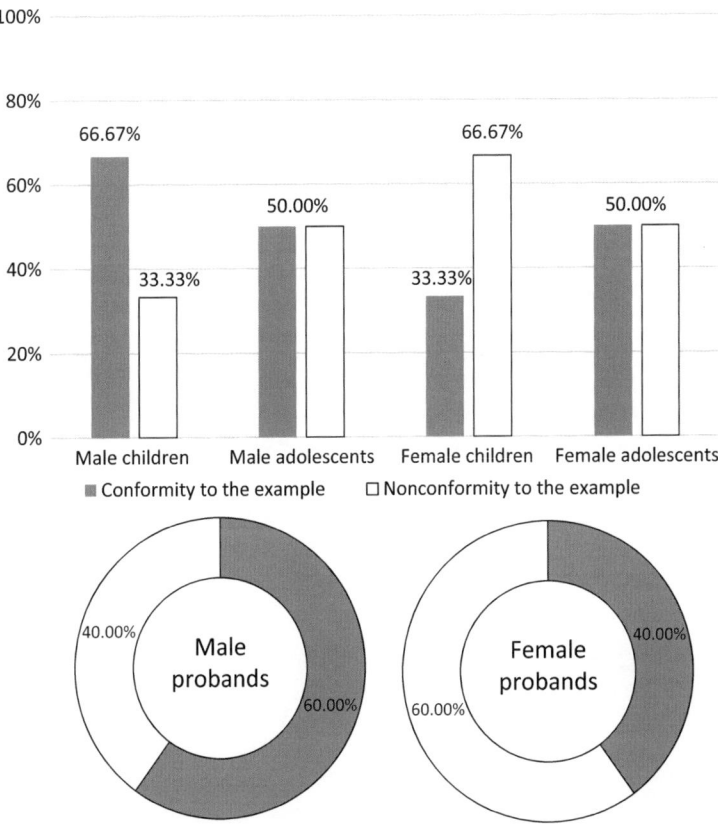

Figure 2.3: Results of the third group (Source: Own illustration).

age groups. Among the adolescents, there are no disparities in the influenceability of the participants. Half of the participants were influenced by the hints in both genders. About 67% of the younger boys were influenced by the sample, whereas only 33% of the younger girls were impacted. Many probands worked only with the picture on the cover and just a few utilized the manual. The gender comparison shows that 40% of the females and 60% of the males were influenced. In all experiments with the pictorial hint, only the correct gender box or the neutral one had an impact on the test subjects.

The outcomes of the fourth experiment confirm the results of the previous two. There are strong differences between the genders and the ages of the test subjects. Only one-third of the younger girls were influenced by one of the hints and none of the older girls. Half of the older male participants were influenced but nearly 90% of the younger boys were impacted by one of the hints. On close examination, the 90% can be split into spoken and visual examples. This shows that more than 70% were influenced by the pictorial sample. Especially the dinosaur and the farm animals

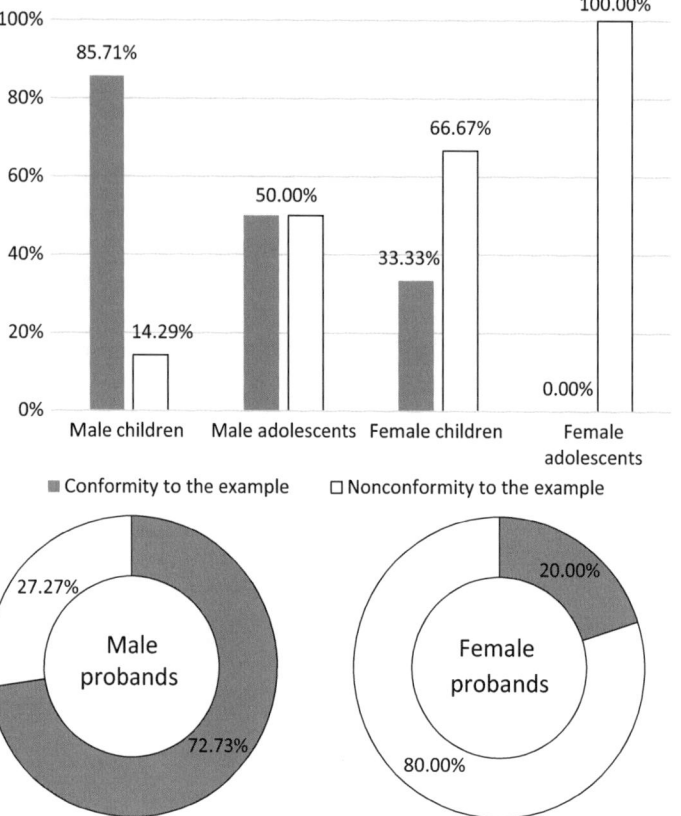

Figure 2.4: Results of the fourth group (Source: Own illustration).

were crafted by the participants quite often. About 70% of the males were impacted by the given hints, but only 20% of the females.

Figure 2.5 shows the overall results of how many test subjects of each gender were influenced by the fixation effect. More than half of the males were influenced by the fixation effect while only one-third of the females were impacted. The younger children of both genders were more influenced than the adolescents. It should be emphasized that the female children are not influenced as much as the male adolescents. The most influenced group are the younger boys followed by the male adolescents and the younger girls. The least influenced group are the female adolescents.

The interesting result that the women, especially in experiment four, are less influenced by the given hints is based on another kind of fixation effect. About 30% of all the females indicated in the questionnaire that they answered subsequent to the experiment, to be influenced by the color of the modeling clay. Almost 40% of

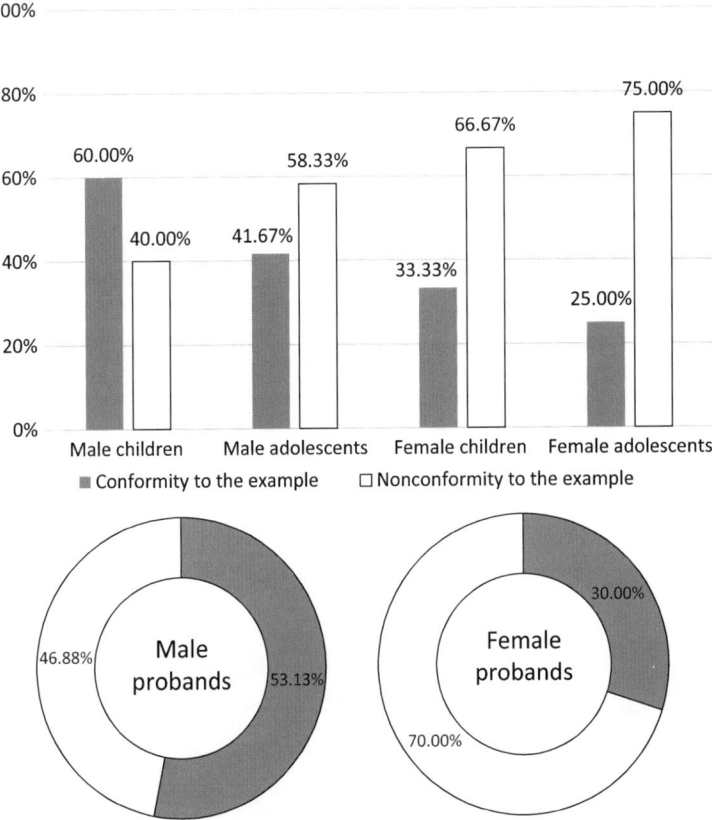

Figure 2.5: General overview (Source: Own illustration).

the female adolescents were governed by the colors. Female children were more influenced by the colors than the adolescent males.

This color fixation shows exactly the opposite results in comparison to the general outcome of the experiments. If this specific fixation effect is also added to the results, over the half of all test subjects were influenced by the examples given in the experiment.

5 Discussion

The literature provides hypotheses about the experiment's findings. The study highlights a number of factors that are emphasized in the literature. The first hypothesis is that men are more affected by the fixation effect than women. In the literature, there are several experiments, e.g. by Jansson and Smith (1991) or Condoor and

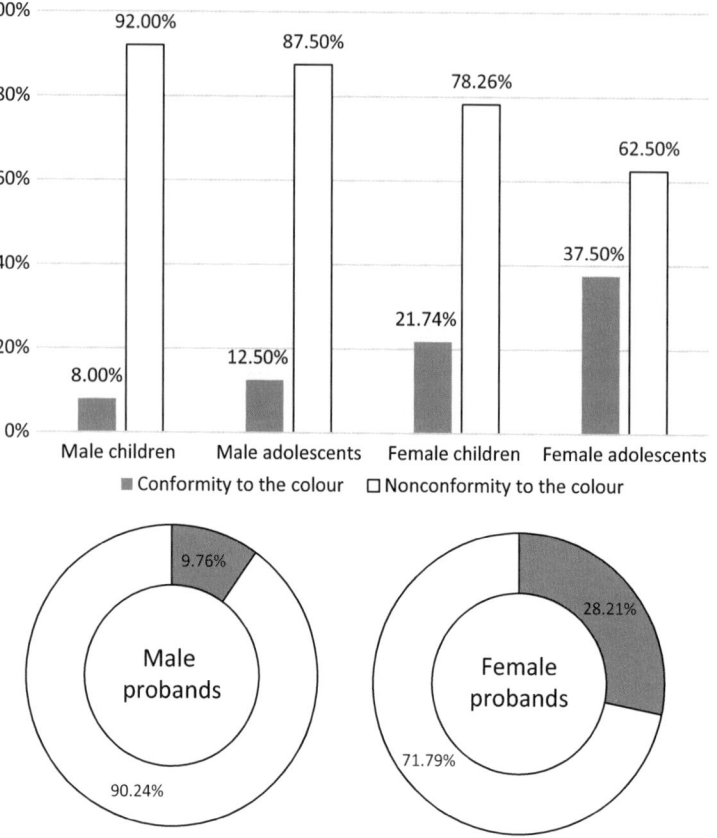

Figure 2.6: Results according to color (Source: Own illustration).

LaVoie (2007), that are explained in Section 2.2 and prove the existence of the fixation effect. The general results of the performed crafting experiment show that the male probands were more influenced by the given hints than the female ones. If the results are considered individually, this hypothesis is only in the first experiment not confirmed, because boys and girls achieved similar results. It is correct to estimate, that men are more affected by the fixation effect, at least with pictures or visual samples like in the second and third experiment. This behavior leads to the question of why women are less influenceable. Women have more experience with handicrafts from childhood on and thus had more time to build up confidence in the execution of such tasks. In gender literature, cognitive differences are explained by biological and sociocultural aspects. Lezak, Howieson, Loring and Fischer (2004) describe that women are better with fine motor skills and are therefore more confident. This could lead to less influenceability when they are crafting because women have more confidence and experience.

The second hypothesis is that visual examples influence both men and women more than verbal examples. Based on the experiments of Goldschmidt and Smolkov (2006) and Chrysikou and Weisberg (2005), it is expected that the visual example influences men and women more than the verbal example. As anticipated, the results of the third and the fourth group of the experiment show a higher amount on influenceability. The impact of the spoken sample was only half that of the visual one. One reason for this could be the duration of the presentation of the hint. The visual sample was present the whole time, whereas the spoken one was only there in the beginning. According to Smith et al. (1993), it makes a difference at which point and how long the hint is present. In order to gain more solid evidence for the fixation effect with only a verbal hint, the experiment needs to be repeated, and the acoustic hint should be present all the time, not only in the beginning.

The third hypothesis is that adolescents, no matter what gender, are less influenceable than children. Agogué and Cassotti (2013) used the 'egg-task' to point out, that adolescents can find better and more solutions for a given problem because children focus too much on existing solutions. Children do not reflect upon the solution as much as adolescents. The instructions and pictures on the box present the perfect solution, which most of the children will start to mimic. However, this gave the probands motivation to create something as beautiful as the picture on the box which enabled them to craft something better than they would have done just from their minds. They occupied themselves a lot more with the topic from the theme box. The theme boxes are aimed at a younger target market, so the stimulus for the younger children was higher than for the adolescents.

The fixation effect is not only limited to solutions, it also can relate to colors. The female adolescents in particular indicate that their handicraft was influenced by the clay's color. In contrast to that, only 8% of the younger boys report that their handicraft was influenced by the color. In the literature, the impact of 'previous knowledge' on the conducted experiments is described. Experience influences individuals in the way they act. This is vastly defined as functional fixedness. (Adamson, 1952) The real, natural color of an existing object is a form of empirical knowledge. People collect this empirical knowledge during their lifetime, therefore this impact becomes stronger with increasing age. Female children are more influenced by the color than the male adolescents. Thus, it is possible that there is a gender difference. Finally, a trend is discernible, that women are more influenced by the color. This topic was not part of the experiments or this chapter but is an interesting aspect for future research.

The experiments confirm the hypothesis because the results are as expected. A clear statement about the fixation effect can be made. It exits in all groups, independent of age or gender. Finally, it needs to be pointed out that the focus of this work is the influenceability by different samples. This research thus does not cover the influenceability in relation to creativity. This means that high levels of influenceability do not automatically imply that an individual is not creative.

The results show that men are more influenced by the fixation effect than women, but nearly everybody is subjected to it. The experiments revealed that the fixation effect only worked with the gender-related theme boxes. The acceptance of the gendered colors was bigger than for the whole gendered box. Nobody who got the opposite theme box was influenced by the visual fixation effect. Eventually, the question remains whether the fixation effect is something positive or negative. This is difficult to answer because, depending on the definition, the evaluation varies, too. Some definitions state that this behavior is blind or inadvertent. The fixation effect limits the observation horizon to a smaller range which restricts the number of solutions to a few that are close to the existing one. This means that people are not able to find completely different ideas anymore or to look at problems and solutions from a whole different perspective. This behavior could be viewed as being bad. On the other hand, there are definitions that present fixation as something positive and encourage a deep exploration of an existing set of solutions. People can find better ways because they are taught to improve the already existing ones. Crilly (2015, p. 15) describes the benefits of repeating solution features as follows: "[...] especially in craft-based design practices, the reuse of good, standard product features is often acceptable or preferable because knowledge of how to shape the product is embedded in prior examples".

6 Conclusion

6.1 Summary

The focus of this work was on the fixation effect and its potential impact on the creativity of men and women. A literature analysis has revealed that there are only minor to no disparities in creativity between males and females. It has also been found that gender differences have not yet been examined in terms of the fixation effect. In order to answer the research question whether there are differences between women and men when exposed to potential solutions to a creative task leading to the fixation effect, an empirical study was carried out.

An experiment was conducted in which children and adolescents between the ages of 6 to 20 years were asked to solve a task silently and individually. The assignment was to create something out of a certain amount of modeling clay. The test subjects were divided into four groups which were confronted with different visual and verbal examples during the 30 minutes processing time. Subsequently, the creations of the test persons were examined and it was evaluated whether a conformity to the given stimuli can be observed.

The results support all three hypotheses and indicate that gender differences in the fixation effect exist. Firstly, it has been shown that men tend to be more attached to examples than women and are influenced more often in their creativity by the presentation of possible solutions. This can be explained by the confidence of women in doing handicrafts. Women are better at fine motoric work; they feel safe and do not need very many instructions. Men tend to be weaker in this area and accept instructions more often. If this experiment was carried out with a stereotypical men's task, such as wood or metal forming, the results might be the opposite. Furthermore, the results agree with prior research on the fixation effect, showing that children are more influenced by examples than adolescents during the generation of new solutions within the creative process. This can be attributed to the fact that children tend to reflect less upon the solutions that are provided to them. Finally, the results of the hypothesis test confirm the presumed differing influence of visual and verbal stimuli on creativity with regards to the fixation effect. The assertions of previous authors have been proven that visual examples are more effective triggers for inspiring solutions than pure verbal examples among both genders. This may be caused by the duration of presenting the hint. The visual sample was present the whole time while the verbal hint was only mentioned in the beginning.

In addition, an unexpected finding was noticed while examining the answers of the test subjects given in a questionnaire after the experiment concerning their source of inspiration. Independently of each other, several participants explained that what they had crafted was influenced by the color of the modeling clay. This interesting outcome can be attributed to the context of functional fixation which limits a person to use an object only in the way it is traditionally used (Duncker & Lees, 1945). In this case, a person is mentally restricted to the given colors and cannot move past the perceptual properties of things that are linked to this certain color scheme.

The fixation effect is often seen in a negative light because it increases the possibility that useful ideas might be overlooked. Nevertheless, fixation is not to be regarded as fundamentally bad or disadvantageous. As a mental obstacle that can make problem-solving more difficult, it is obviously possible that fixation prevents good, appropriate solutions from being considered. But just as there may be pragmatic advantages for a deep exploration of a narrow set of viable solutions, creative benefits and time saving might occur as well (Crilly & Cardoso, 2017). However, in the global business environment, the problem of developing new products seldom comes with example solutions. Instead, innovative companies often subconsciously look to other devices, models or examples that they have encountered or may encounter while working on a problem, on which they could also become fixated.

With the experiment in this work, previous research on the fixation effect with its obvious interdisciplinary character that links the technologically complex discipline of engineering design and the quite distinct nature of cognitive sciences is supported by new results. Furthermore, this study closes a gap in interdisciplinary research on creativity, by focusing on gender differences within the fixation effect. Thus, this chapter contributes to the vast body of literature on gender differences in creativity.

6.2 Limitations

Even though the experiment described in this chapter was conducted in a systematic and robust way, there are several limitations that need to be mentioned, as well as possibilities for future research directions.

Due to the quasi-experimental study design, a lack of randomization might have introduced some bias to the results. The predetermined adolescent participants were students of the same college of further education in Erlangen and all younger participants were children of employees and students of the Friedrich-Alexander-University Erlangen-Nuremberg and the University Hospital, the Max Planck Institute, the Fraunhofer Institute and the Erlangen-Nuremberg Student Union. This might limit the generalizability of the conclusive results to a small part of the population and reduce internal validity. In the future, the presented study might be complemented by more complex designs using an extended sample size to give a better representation of the population as a whole.

Although the task of the experiment was administered silently and individually, a completely distraction-free environment could not be achieved at all time. The spatial conditions for the execution of the experiment with the adolescents were limited. Since the provided space had to be shared with other scientists who conducted different experiments with the participants, the area in which the subjects performed the task described in this chapter could only be separated by moveable partitions. This occasionally resulted in a very noisy environment during the experiment, which may have caused distractions among the participants and may have unintentionally influenced their creativity. During the subsequent experiment with the younger participants, this problematic issue was addressed by providing a separate room, in which the children could complete the described task. This prevented any distractions caused by external impulses from the outset. Future work needs to take this into consideration as well when conducting further studies.

All adolescents and children should be completely unbiased and uninformed about the task and the materials used in the experiment so that they would not have the chance to practice handicrafts or be influenced by factors other than those given during the experiment. However, the exchange between individual participants of different experimental groups could not be completely prevented. In addition, the experiment was inadvertently announced in the 'FAU Pfingstferienbetreuung', under the heading 'Crafting with Fimo'. Thus, a bias of some participants is not completely excludable. Future studies could try to eliminate any potentially biased opinions as much as possible.

As explained in the second section of this chapter, a non-complex problem was chosen that does not require any technical nor scientific knowledge or skills to solve it. The creative task was supposed to be fulfilled equally well by both age groups in order to ensure the comparability of the results. The modeling clay Fimo with its sets was ideal for this purpose. Nevertheless, it must be emphasized that adolescents may not have been influenced by the pictorial samples to the same extent as the children, since the Fimo kids form&play sets address a younger target group. It is possible that the percentage of adults that was influenced by the visual examples might have been higher if other incentives had been presented that are more appropriate for this age group. Therefore, it would be interesting for future works to use corresponding stimuli.

Furthermore, it should be noted that an attempt has been made to keep the given influencing information as gender neutral as possible. By utilizing the Fimo in the experiment, it was an obvious choice to include the themed boxes offered by Staedtler as stimuli as well. Some of those sets, however, were gendered, which means, that they relate to people of one particular sex more than the other (Morrison & Shaffer, 2003). This is manifested e.g. by the different color combinations offered or that either a crafting boy or girl is shown on the cover of the different boxes. An attempt was made to distribute all gendered and neutral Fimo sets equally between the male and female participants and condition groups in order to minimize any distortions. Future research could address this issue by examining whether men and women are similarly influenced by gendered stimuli.

6.3 Implications for further research

In addition to the implications for future research resulting from the described limitations mentioned in the previous section, the findings of the current study indicate several possible avenues for future research. As mentioned in Section 2.4, it was observed that adults are more influenced by the colors of clay than children. This functional fixation on colors should be further examined in future work to investigate the extent to which men and women differ in this area.

Furthermore, based on the results of this work, the research of Agogué et al. (2013) on the varying nature of examples should be extended. The authors have demonstrated with the C-K theory that is possible to 'reinforce' or 'overcome' fixation by providing ideating people with relevant stimuli. Some of the examples were purposefully selected to be 'fixating' and others to be 'de-fixating'. The C-K theory was used to generate the referential to assess the fixation and to generate 'fixating' and 'de-fixating' stimuli. These aspects were deliberately not considered in the present study, as this chapter represents a first step to address gender differences within the field of fixation. Therefore, it would be interesting to examine in future works whether stimuli that may either induce fixation (restrictive example) or help to overcome it, i. e. 'de-fixating' (expansive examples) influence males and females to the same extent.

6.4 Implications for practice

In addition to the scientific significance and the indications for future research, several practical implications can also be derived from this work. The results of this experiment could assist in the development of tools and techniques for managers to clarify the nature of the cognitive biases in creativity and to mitigate fixation in practice. The generation of novel ideas is essential to any innovative endeavor in a company. Yet one of the main obstacles to creativity is the fixation effect (Ezzat, Agogué, Le Masson, & Weil, 2017). Instructions from leaders can have a significant impact on the ideation ability of their followers. Therefore, managers could adapt their instructions during a creative idea generation task of their subordinates accordingly.

Furthermore, the results can be applied to marketing activities, especially the packaging of products. Since men are more likely to be influenced by examples, the depiction of possible solutions for the use of a product on the packaging might be utilized to market supposedly gendered products to a larger target audience. For instance, the modeling clay Fimo from the company Staedtler, which was used as the material for the experiment conducted in this work, is primarily regarded as a product for women and children. By providing examples on the packaging, it could become attractive to men as well and thus expand the potential target market.

References

Adamson, R. E. (1952). Functional fixedness as related to problem solving: a repetition of three experiments. *Journal of Experimental Psychology*, 44(4), 288–291.

Adamson, R. E., & Taylor, D. W. (1954). Functional fixedness as related to elapsed time and to set. *Journal of Experimental Psychology*, 47(2), 122.

Agogué, M., & Cassotti, M. (2013). Understanding fixation effects in creativity: A design-theory approach. In *DS 75-2: Proceedings of the 19th International Conference on Engineering Design (ICED13). Design for Harmonies. Vol.2: Design Theory and Research Methodology* (pp. 103–112).

Agogué, M., Kazakçi, A., Hatchuel, A., Masson, P., Weil, B., Poirel, N., & Cassotti, M. (2013). The Impact of Type of Examples on Originality: Explaining Fixation and Stimulation Effects. *The Journal of Creative Behavior*, 48(1), 1–12.

Agogué, M., Poirel, N., Pineau, A., Houdé, O., & Cassotti, M. (2014). The impact of age and training on creativity: A design-theory approach to study fixation effects. *Thinking Skills and Creativity*, 11, 33–41.

Amabile, T. M. (1983). The social psychology of creativity: A componential conceptualization. *Journal of personality and social psychology*, 45(2), 357.

Amabile, T. M., & Khaire, M. (2008). Your organization could use a bigger dose of creativity. *Harvard Business Review*, 86(10), 101–109.

Anderson, N., Dreu, C. K. W. de, & Nijstad, B. A. (2004). The routinization of innovation research: A constructively critical review of the state-of-the-science. *Journal of organizational Behavior*, 25(2), 147–173.

Anderson, N., Potočnik, K., & Zhou, J. (2014). Innovation and creativity in organizations: A state-of-the-science review, prospective commentary, and guiding framework. *Journal of management*, *40*(5), 1297–1333.

Baer, J., & Kaufman, J. C. (2008). Gender Differences in Creativity. *The Journal of Creative Behavior*, *42*(2), 75–105.

Bender, S. W., Nibbelink, B., Towner-Thyrum, E., & Vredenburg, D. (2013). Defining characteristics of creative women. *Creativity Research Journal*, *25*(1), 38–47.

Bonnardel, N., & Marmèche, E. (2004). Evocation Processes by Novice and Expert Designers: Towards Stimulating Analogical Thinking. *Creativity and Innovation Management*, *13*(3), 176–186.

Cheung, P. C., & Lau, S. (2010). Gender differences in the creativity of Hong Kong school children: Comparison by using the new electronic Wallach-Kogan creativity tests. *Creativity Research Journal*, *22*(2), 194–199.

Chrysikou, E. G., & Weisberg, R. W. (2005). Following the wrong footsteps: fixation effects of pictorial examples in a design problem-solving task. *Journal of Experimental Psychology: Learning, Memory, and Cognition*, *31*(5), 1134.

Condoor, S., & LaVoie, D. (2007). Design Fixation: A Cognitive Model. In *Proceedings of ICED 2007, the 16th International Conference on Engineering Design*, 28–31.

Cooper, R. G. (1988). Predevelopment activities determine new product success. *Industrial Marketing Management*, *17*(3), 237–247.

Crilly, N. (2015). Fixation and creativity in concept development: The attitudes and practices of expert designers. *Design Studies*, *38*, 54–91.

Crilly, N., & Cardoso, C. (2017). Where next for research on fixation, inspiration and creativity in design? *Design Studies*, *50*, 1–38.

Csikszentmihalyi, M. (1999). 16 implications of a systems perspective for the study of creativity. In R. J. Sternberg (Ed.), *Handbook of creativity* (pp. 313–335). Cambridge, MA, US: Cambridge University Press.

Duncker, K., & Lees, L. S. (1945). On problem-solving. *Psychological Monographs*, *58*(5).

Ezzat, H., Agogué, M., Le Masson, P., & Weil, B. (2017). Solution-oriented versus Novelty-oriented Leadership Instructions: Cognitive Effect on Creative Ideation. In J. S. Gero (Ed.), *Design Computing and Cognition'16* (pp. 99–114). Cham: Springer.

Goel, V., & Pirolli, P. (1992). The structure of design problem spaces. *Cognitive science*, *16*(3), 395–429.

Goldschmidt, G., & Sever, A. L. (2011). Inspiring design ideas with texts. *Design Studies*, *32*(2), 139–155.

Goldschmidt, G., & Smolkov, M. (2006). Variances in the impact of visual stimuli on design problem solving performance. *Design Studies*, *27*(5), 549–569.

Guindon, R. (1990). Designing the design process: Exploiting opportunistic thoughts. *Human-Computer Interaction*, *5*(2), 305–344.

Hanington, B. (2003). Methods in the making: A perspective on the state of human research in design. *Design issues*, *19*(4), 9–18.

He, W.-j., & Wong, W.-c. (2011). Gender differences in creative thinking revisited: Findings from analysis of variability. *Personality and Individual Differences*, *51*(7), 807–811.

Henard, D. H., & Szymanski, D. M. (2001). Why some new products are more successful than others. *Journal of marketing Research*, *38*(3), 362–375.

Hong, E., & Milgram, R. M. (2010). Creative thinking ability: Domain generality and specificity. *Creativity Research Journal*, *22*(3), 272–287.

Jansson, D. G., & Smith, S. M. (1991). Design fixation. *Design Studies*, *12*(1), 3–11.

Karwowski, M., Lebuda, I., Wisniewska, E., & Gralewski, J. (2013). Big five personality traits as the predictors of creative self-efficacy and creative personal identity: Does gender matter? *The Journal of Creative Behavior*, *47*(3), 215–232.

Kaufman, J. C., Baer, J., Agars, M. D., & Loomis, D. (2010). Creativity stereotypes and the consensual assessment technique. *Creativity Research Journal, 22*(2), 200–205.

Lezak, M. D., Howieson, D. B., Loring, D. W., & Fischer, J. S. (2004). *Neuropsychological assessment:* Oxford University Press, USA.

Luchins, A. S. (1942). Mechanization in problem solving: The effect of Einstellung. *Psychological Monographs, 54*(6).

Luchins, A. S., & Luchins, E. H. (1969). Einstellung effect and group problem solving. *The Journal of Social Psychology, 77*(1), 79–89.

Marsh, R. L., Landau, J. D., & Hicks, J. L. (1996). How examples may (and may not) constrain creativity. *Memory & Cognition, 24*(5), 669–680.

Marsh, R. L., Ward, T. B., & Landau, J. D. (1999). The inadvertent use of prior knowledge in a generative cognitive task. *Memory & Cognition, 27*(1), 94–105.

Morrison, M. M., & Shaffer, D. R. (2003). Gender-role congruence and self-referencing as determinants of advertising effectiveness. *Sex Roles, 49*(5–6), 265–275.

Purcell, A. T., & Gero, J. S. (1996). Design and other types of fixation. *Design Studies, 17*(4), 363–383.

Rhodes, M. (1961). An analysis of creativity. *The Phi Delta Kappan, 42*(7), 305–310.

Runco, M. A., & Jaeger, G. J. (2012). The standard definition of creativity. *Creativity Research Journal, 24*(1), 92–96.

Sarkar, P., & Chakrabarti, A. (2008). The effect of representation of triggers on design outcomes. *AI EDAM (Artificial Intelligence for Engineering Design, Analysis and Manufacturing), 22*(2), 101–116.

Sayed, E. M., & Mohamed, A. H. H. (2013). Gender differences in divergent thinking: use of the test of creative thinking-drawing production on an Egyptian sample. *Creativity Research Journal, 25*(2), 222–227.

Schulthess, M. (2012). *Die Nutzung von Analogien im Innovationsprozess.* Wiesbaden: Gabler Verlag.

Simon, H. A. (1996). *The sciences of the artificial.* Cambridge, MA, US: MIT press.

Smith, S. M., & Blankenship, S. E. (1991). Incubation and the persistence of fixation in problem solving. *The American journal of psychology,* 61–87.

Smith, S. M., Ward, T. B., & Schumacher, J. S. (1993). Constraining effects of examples in a creative generation task. *Memory & Cognition, 21*(6), 837–845.

Stein, M. I. (1953). Creativity and culture. *The journal of psychology, 36*(2), 311–322.

Steiner, G. (2011). *Das Planetenmodell der kollaborativen Kreativität.* Wiesbaden: Gabler.

Sternberg, R. J., & Lubart, T. I. (1999). The concept of creativity: Prospects and paradigms. In R. J. Sternberg (Ed.), *Handbook of creativity* (pp. 3–15). Cambridge, MA, US: Cambridge University Press.

Stoltzfus, G., Nibbelink, B. L., Vredenburg, D., & Hyrum, E. (2011). Gender, gender role, and creativity. *Social Behavior and Personality: an international journal, 39*(3), 425–432.

Taylor, C. W. (1988). Various approaches to and definitions of creativity. *The nature of creativity,* 99–121.

Thompson, L. (2003). Improving the creativity of organizational work groups. *Academy of Management Perspectives, 17*(1), 96–109.

Utterback, J. (1994). *Mastering the dynamics of innovation: How companies can seize opportunities in the face of technological change.* Boston, MA: Harvard Business School Press.

Vasconcelos, L. A., & Crilly, N. (2016). Inspiration and fixation: Questions, methods, findings, and challenges. *Design Studies, 42,* 1–32.

Philipp Hässler, Sven Schneider
3 Effects of Gender Differences in Competition on Creativity

1 Introduction

The topic of gender differences in the work environment or even in people's private lives is an up-to-date topic which is often featured in the news and reflected in several movements like HeForShe or predetermined women quota in companies. The cause for this lies in large differences between men and women in professional life. As a matter of fact, from 1992 to 1997, only 2.5% of the top executives in US firms were women (Bertrand & Hallock, 2001). Great gaps still existed even in 2008 with only 2.8% of companies in the S&P 500 having a female Chief Executive Officer (CEO). Traditionally, these differences could be explained by discrimination against women in the workplace, as well as differences in skill and preferences. For example, men might find it easier to learn a specific skill that is needed for management positions quicker than women and women do not enjoy working long hours (Niederle & Vesterlund, 2008). Today, however, more women than ever enter the job market and other explanations have been studied.

Differences in competitive behavior have been used to explain the gender gap, especially in competitive fields. Several studies have concluded that men are more inclined to compete and that they perform better, even when no prior performance difference exists in a non-competitive environment (Gneezy, Niederle & Rustichini, 2003).

It is equally important that the industry today is shifting from traditional, straight forward methods to methods like Design Thinking and entrepreneurship, where creativity is increasingly viewed as an important asset of every company that wants to succeed (Dimitriadis, Anastasiades, Karagiannidou & Lagaki, 2018). Literature suggests that women perform equally well, if not better, in creative tasks than men.

Gender differences in both creativity and competition have been analyzed before. However, effects in competitive behavior have not been linked to creativity yet. Due to recent developments, the importance of creativity and the nature of competitive fields have been highlighted. Thus, our goal with this study is to analyze, how the response to competition affects creativity in women and men.

For this purpose, we designed two experiments, one with a non-competitive piece rate payment and a competitive 'winner-takes-it-all' payment. Seventeen boys and seventeen girls were tasked with generating as many alternative uses for safety glasses by a well-known company as possible. This relates back to divergent thinking, which is a part of creativity and is widely used for assessing creativity.

This chapter is structured as follows. First, we outline the theoretical basis for understanding the context of the experiment. Then we explain the methodology in detail and finally, we lay out our results, discuss our findings and conclude our chapter with given research implications and limitations.

2 Theoretical Basis

The following section will illustrate and give deeper insights into the context of competition and creativity by connecting available literature on the topic. Firstly, we examine the topic of creativity, its importance in innovation and differences in the creative behaviour of men and women. In the second section, we will analyze the response to competition of women and men. In a third step, we aim to connect both fields and establish our research hypotheses based on existing literature.

For this purpose, a systematic literature review underlies this study. We searched the data-bases Scopus, Google Scholar and Business Source Complete using the keywords 'competition', 'creativity', 'gender difference', 'gender', as well as 'divergent thinking'. Relevant combinations of the keywords were considered as well. Our literature was selected by taking the quality of the journal, the relevance of the keywords, the abstract and conclusions into account.

2.1 Creativity, Divergent Thinking and Gender

Following the definition from Runco and Jaeger (2012), a creative product or idea is original and effective at the same time. In this regard, the degree of originality is determined by novelty, uniqueness or statistical rarity. Effectiveness is measured by looking at the functionality and usefulness of the response (Abraham, 2016).

The alternate uses task is supposed to determine both originality and effectiveness. In this context, participants are asked to list alternative uses of a product. Answers that are not named as frequently are thus more creative. However, they must be relevant to the topic or else they are not considered to be creative (Abraham, 2016). The alternate uses task is categorized as a 'Divergent Thinking' task, which is a widely used way of measuring creativity.

'Divergent Thinking' is a way of finding as many solutions as possible from a single starting point. However, it is not to be confused with 'convergent thinking' which aims to find exactly one solution starting from multiple points (Brophy, 2001).

Nowadays, new developments take place at a rapid pace and businesses need to adapt if they want to prosper and succeed in their respective field (Proctor, 1991). Within well-run companies, creativity and innovation have been identified as a sure means to respond to changes in the market or the business itself (Dimitriadis et al., 2018).

In the context of divergent thinking, creativity is the ability to come up with many new solutions, while innovation is considered the implementation of creativity (Sokolova, 2015). Moreover, creativity is viewed as an important aspect of managers' work. They need to be able to adapt to an increasing number of problems to which no prior solutions exist (Ackoff & Vegara, 1988). Creativity is used to solve existing problems in new ways or to solve new problems. Furthermore, cost-effective solutions require management to encourage their employees to think outside of the box and provide them with the methods and resources they require (Dimitriadis et al., 2018). Thus, it can be said that creativity in connection with innovation has always been an important aspect for the success of businesses. Innovative entrepreneurs are also essential to developed economies because they support their continued dominance (Adams, 2005).

Research in the field of gender differences in creativity has found conflicting notions, with some studies describing women as more creative and others men. However, the studies that reported no or only small differences between the genders are more numerous than those that reported the opposite (Abraham, 2015). Nonetheless, girls or women slightly outperform boys or men when it comes to creativity (Baer and Kaufman, 2008). Additionally, women perform better in most tests when the ability of divergent thinking is assessed as a creative ability (Lin, Hsu, Chen & Wang, 2012).

External factors without a doubt influence the creative potential of males and females differently. For example, it is suggested that extrinsic motivators like rewards, money or evaluations negatively impact the creativity of girls. However, the same cannot be stated for boys (Deci, Cassio & Krussel, 1973). Koestner, Zuckerman and Koestner (1989) showed that praise could sometimes positively impact boys but always had a negative influence on girls when it was used as a motivator.

According to the available literature, women should be equally, if not more accomplished than men in creative disciplines. However, men show a much larger creative productivity than women. This means that men find more success than women in fields like literature, art and design. To some extent, this effect can already be explained by expectations that are placed on women in society, which include raising a family, as well as the lack of different opportunities due to discrimination (Baer & Kaufman, 2008). Nonetheless, it is interesting to contemplate, why the differences are that large. Another possible explanation could be different responses to competition which will be explored in the next section of this study. The following section will address differences in competitive behavior in men and women and outline the consequences.

2.2 Differences in Competition

A study by Bertrand and Hallock (2001), from 1992 to 1997 found that only 2.5% of the five highest-paid top-executives from several American companies are women.

According to literature, this phenomenon could be explained by occupational self-selection, discrimination in hiring and promotion, as well as a lack of long-term commitment from women (Blau & Kahn, 2000). While a large gender gap persists in the unequal distribution of top-management and wage, it must be noted that the gender pay gap is continuously decreasing in the long term (Blau & Kahn, 2000). Several works have touched on different subjects while aiming to explain this gender gap. So far, research was rather focused on gender differences in negotiating (Babcock & Laschever, 2003), overconfidence (Barber & Odean, 2000) and, finally, differences in the response to competition (Gneezy, Niederle & Rustichini, 2003; Price, 2008; Niederle & Vesterlund, 2008). Particularly interesting for this chapter is the last example. Competition plays a vital role in business positions, politics, law and top-management. These fields are defined by strong forms of competition because of their high correlation with wealth and power (Price, 2008). In general, competition is a good means of increasing performance. For example, many institutions provide scholarships and prizes on a competitive basis (Price, 2008).

Several studies imply a gender gap in performance when it comes to competitive tasks. Gneezy et al. (2003) analyzed the behavior of men and women under a non-competitive piece rate payment and a competitive 'winner takes it all' payment scheme and concluded that men significantly increase their performance in mixed gender competitions. In addition, a study observed boys and girls when competing against each other in running and came to a similar conclusion (Gneezy & Rustichini, 2004). Consequently, it can be said that gender differences do exist in competitive environments. A reason for this could be that women do not attribute success to inner traits like skill quickly, while the opposite is true for men (Beyer, 1990).

Differences also persist when examining the gender composition of the groups. It was found that women perform significantly better in single-sex groups than in mixed groups, while the performance of men stays the same in both conditions (Gneezy et al., 2003).

Niederle and Vesterlund (2007) observed differences in the willingness to enter competitions. They created circumstances where men and women performed similarly. Regardless of this fact, men were more than two times more likely to enter competitions than women. They go on to explain the gender gap in competition entry as a result of overconfidence of men, as well as different preferences regarding the participation in competition (Niederle & Vesterlund, 2007). In a later study, they found that men prefer to be compensated in a tournament scheme, while women prefer piece-rate payments (Niederle & Vesterlund, 2011). Overall, this means that fewer women than men compete in tournaments and consequently, fewer women win tournaments. Given these facts, women are less likely to acquire promotions or better jobs through competing in contests (Niederle & Vesterlund, 2007).

As can be seen, there are large gender gaps in individual competitions in mixed groups. However, as mentioned above, women perform better in single-sex groups and are thus able to close the gender gap. Other circumstances that show differing effects in competitiveness in men and women include competing in a team, competing against one's own score and facing competition that is perceived to be better than oneself.

Healy and Pate (2011) observe gender differences when competing in a team and find that women are more willing to compete in a team while men rather want to compete on their own. They also found that the gender gap is, in turn, decreased by almost two thirds when competing in a team. In their study, they link this phenomenon to competitive preferences in men and women (Healy & Pate, 2011).

Another study found that when competing against one's previous performance, no gender gap exists and that competing against oneself is as good as other competition for boosting performance (Apicella, Demiral & Mollerstrom, 2017). In that regard, it is suggested that a shift to self-competition in labor and economic markets could add to reducing the gender gap. Interestingly, gender differences undergo a change when facing competition that is perceived as superior. John (2017) found that the performance of men decreased while women's performance did not change when facing the top of the class in comparison to facing the bottom of the class.

3 Research Hypotheses

In order to answer the research question, three research hypotheses from the theoretical background are going to be established in the following. From the literature, we know that there seems to be a high similarity between women and men when it comes to being creative. However, it is generally concluded that women slightly outperform men (Forisha, 1978; Lin et al., 2012). Moreover, it is implicated that women are negatively influenced by extrinsic motivators (Deci et al., 1973). Thus, for our first hypothesis we conclude:

Hypothesis 1: In a non-competitive setting, women tend to generate slightly more creative ideas than men.

Regarding differences in competitiveness, it is known that men increase their performance in a mixed gender competition, while the performance of women stays the same. Assuming this relation can be transferred to creativity, we arrive at:

Hypothesis 2: In a competitive setting, men generate more creative ideas than in a non-competitive setting

We derive our final research hypothesis from a combined analysis of existing literature. Women perform slightly better in creative tasks, while men are superior in mixed gender competition. Thus, we can assume that:

> **Hypothesis 3:** In a competitive setting, the gap between men and women decreases compared to the non-competitive setting where women perform better.

4 Methodology

In the following sections, we will outline the design of the conducted experiment with which the above-developed hypotheses are supposed to be tested.

4.1 Design of Experiment

We based our approach on an experiment conducted by Gneezy et al. (2003) where participants were asked to solve mazes under a non-competitive piece-rate payment, as well as a competitive 'winner takes it all' payment.

In our case, we designed an alternative uses task experiment which is widely used in the context of divergent thinking to measure creativity. Participants were asked to generate as many possible alternative uses for pairs of safety glasses that are produced by a well-known German manufacturer as possible. Like Gneezy et al. (2003), two separate experiments were conducted.

The experiments were conducted in the workshop area of a startup incubator located in Bavaria, Germany in the context of a "Girls' and Boys' Day", together with multiple other experiments. Due to limited space at the provided location, a makeshift room was created with dash panels. Four tables were spread out in a half-open room, with two seats available at each table. The view was obstructed on two sides of the area. For the purpose of the task, four safety glasses were placed on each table for inspiration. The task was performed in mixed gender groups of eight people, when possible. After a short introduction of the items and explanation of the task, the participants were asked to generate ideas. To prevent participants from influencing each other, group work and talking were prohibited during the experiment. The participants had ten minutes to generate as many alternatives as possible in each experiment.

In the first experiment, the participants were paid ten cents per generated alternative use, independent of the other students' ideas. In the second experiment, only the participant who generated the most ideas was given a 10 € gift voucher. Even though non-monetary incentives like praise also often work well as a motivator, we decided to utilize monetary rewards because their effectiveness depends less on exterior influences (Kosfeld & Neckermann, 2011), contrary to non-monetary incentives (for example what the task means for the user personally). Furthermore, Gneezy et al. (2003) had already performed a similar experiment with monetary incentives.

Twelve students, six women and six men, participated in the first experiment, while 22 students, eleven men and eleven women, performed the task in the competitive setting.

4.2 Approach to Evaluation

We analyzed our experiment according to the aspects of relevance and originality, which is one way to define a creative idea. Thus, we first filtered out irrelevant ideas with regards to the task at hand, for example, using the safety glasses in a way that is not feasible. Furthermore, we used the definition of originality as described in the Torrance Test of Creative Thinking (TTCT) (Kim, 2006): Originality represents the number of ideas that are generated infrequently in comparison to other ideas. Taking this into account, we assumed that every idea named by less than 50% of the participants can be viewed as original and thus is an uncommon or unique response. Overall, we designed an analysis sheet, in which all ideas were entered and filtered accordingly. The numbers 1 to 34 each represent one participant. Table 3.1 shows an exemplary layout. It was recorded, how often an idea was mentioned in a competitive and non-competitive setting. The generated ideas and the gender of each participant were registered. We calculated the sum over each idea in the competitive and non-competitive setting. Each number represents one participant. Furthermore, we divided the table by gender for further analysis of the respective differences.

Table 3.1: Excerpt of our evaluation sheet (Source: Own illustration).

Participant	Ideas for alternative use of glasses	Sum	Sum competitive environment	Sum non-competitive environment
1	Protection for handcrafting	...	...	...
2	Sports glasses	...	...	...
3	Ski glasses	5	3	2
...				
...				
34				

5 Results

Based on the evaluation sheets we created graphs that visualize the results of the experiments. Figure 3.1 shows the performance of male participants in the non-competitive setting.

Altogether, the six participating men generated 15 original ideas in the fixed incentive-based experiment, with the highest number of original ideas being four and

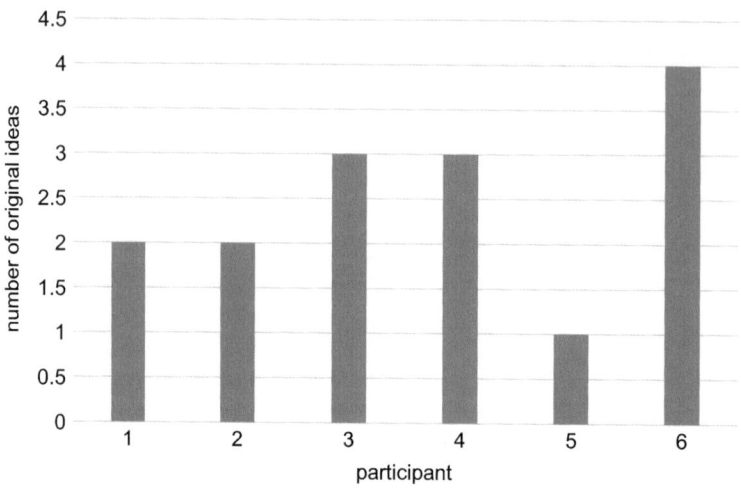

Figure 3.1: Results in a non-competitive setting, men (Source: Own illustration).

the lowest number being one. This means that on average, the participants came up with 2.5 solutions per person. Furthermore, the calculation of the standard deviation in the non-competitive setting for men resulted in 1.05. In comparison, these numbers are slightly lower than the results of their female counterparts. The results of the same experiment conducted with women are depicted in Figure 3.2.

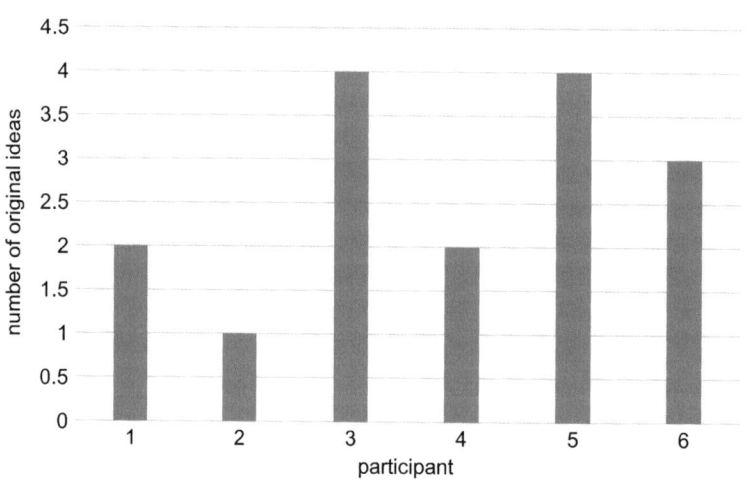

Figure 3.2: Results in a non-competitive setting, women (Source: Own illustration).

The graph indicates that both genders performed nearly identically, with women generating one original idea more with a total of 16. This results in an average of 2.7 ideas per person. The calculation of the standard deviation resulted in 1.34 accordingly. In line with the outcome of the first experiment, the highest number of ideas was four, the minimum one. Due to the small sample size, these results entail inaccuracies. More on the topic of limitations of the experiments can be found in Section 3.7 'Research Limitations and Implications'.

In contrast to the results of the experiment under non-competitive circumstances, the competitive environment yielded far better results. Figure 3.3 depicts that men provided 53 answers overall (the higher number of participants must be considered when comparing the two different experiments).

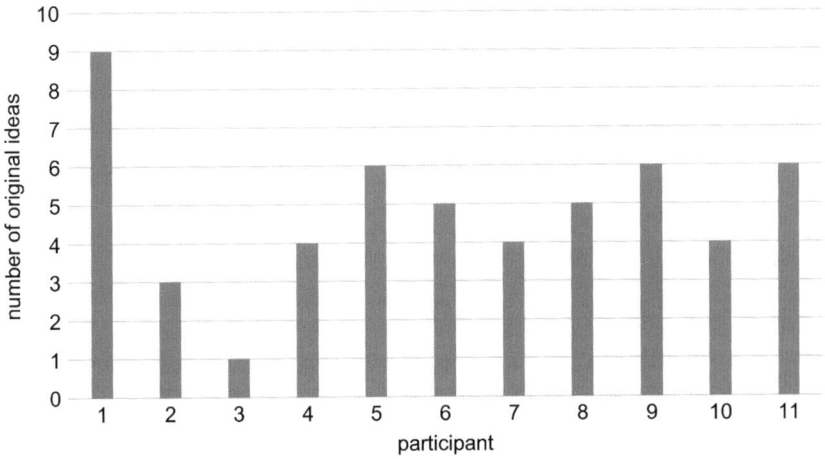

Figure 3.3: Results in a competitive setting, men (Source: Own illustration).

Accordingly, the men averaged at 4.8 original ideas and thus were nearly two times more creative in the competitive environment than in the non-competitive one (2.5 ideas on average). Although the minimum number of solutions stayed the same with one idea (participant 3), the maximum number increased to nine answers. Also, the standard deviation increased to 2.04 as opposed to the previous 1.05.

Figure 3.4 reveals that the female participants did not improve as much as the men in the competitive environment. We obtained 33 answers in total with an increase to 3.0 original ideas per person, which is slightly higher than the 2.7 in the non-competitive setting. Similarly, the standard deviation followed the same trend as observed for the men. It increased from 1.34 to 2.32. Furthermore, the illustration also shows that two participants failed to provide even a single original solution. Nevertheless, the maximum number of ideas nearly doubled to seven. Thus, the differences in performance between individual female participants increased considerably

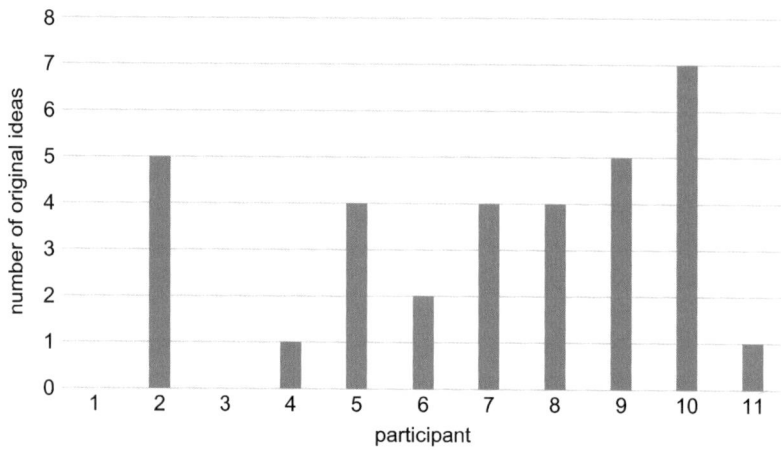

Figure 3.4: Results in a competitive setting, women (Source: Own illustration).

in a competitive setting compared to the non-competitive setting. A setting dependent difference in individual performance was less observable for male participants.

6 Discussion

The following section aims to compare our results to the theoretical part of this chapter. Overall, the goal of this study was to analyze gender-specific effects of competition on creativity. We developed research hypotheses in Section 3.2. and we will discuss their validity in view of the obtained experimental results in the following.

For our first hypothesis, we concluded that women should generate slightly more creative ideas than men in a non-competitive setting. This goes back to the theory that states that girls or women outperform boys and men by a small margin regarding creativity (Baer & Kaufman, 2008). This should especially be the case when it comes to divergent thinking capabilities, which the alternative uses tasks are a part of (Lin et al., 2012). The results of 2.7 average ideas per women and 2.5 per men seem to be in line with this theory since women in our experiment also generated slightly more ideas than men. However, the differences are marginal and thus this work can be categorized as contributing to the research that indicates no overall difference. Moreover, the literature states that extrinsic motivators, e. g. money, negatively impact girls but not boys (Deci et al., 1973). In our case, we could not observe any negative impact of the fixed 10 ct incentive pay, which we used as a motivator. This could mean, that the extrinsic motivator was either not high enough or the participants simply were not impacted. Furthermore, our results showed that the standard deviation for women is slightly higher than for men. This could relate back to our group composition, which could have included a greater variety of highly-creative and non-creative women in our case, thus resulting in a higher standard deviation than observed for the men. Otherwise, our results indicate that women and men start off on an equal footing when it comes to creativity in non-pressure/competitive settings. Of course, in our case, the sample size is too small to generalize any conclusions.

With our second hypothesis, we suggested that men generate more creative ideas in a competitive setting compared to a non-competitive one. The basis of this hypothesis is that men increase their performance drastically in competition with mixed genders (Gneezy et al., 2003) and that this behavior is not influenced by the nature of the competition itself (Gneezy & Rustichini, 2004). Therefore, the execution of a creativity contest should yield a comparable outcome. In the conducted experiment, men generated an average of 2.5 ideas in a non-competitive setting. This number drastically increased to an average of 4.8 when the participants competed against each other. Thus, the results support the hypothesis that men generate more creative ideas in this type of setting. The extent of the increase in performance in the experiment may be amplified by several factors that were analyzed by other researchers. This includes the fact that men prefer to compete on their own (Healy and Pate, 2011). Furthermore, another supporting factor is that the students did not face competitors which may be perceived as superior, due to them only competing against other participants of their own age and educational background. Literature indicates that facing competition that is regarded as superior is detrimental to male performance (John, 2017). Overall, the degree of the increase in performance could partially be due to ideal pressure conditions for the men. The increase in performance of almost 100% could also relate back to having better performing male students in the competitive setting. Similar to the first hypothesis, where we discussed that the monetary incentives did not impact males or females, we can derive that the non-competitive circumstances were not ideal for boys and therefore the impact of the competitive environment is as large as it is. We have also observed a higher standard deviation in the competitive environment. This could be the result of men having different attitudes towards competition which implies that our group consisted of highly competitive men as well as men who do not prefer competition. This observance is also supported by the theory that while men generally prefer to compete, of course not all of them do (Niederle & Vesterlund, 2007), thus increasing the gap between them. Again, we must note that our sample size was too small and could impact standard deviations, because our sample could have had many over-performing or underperforming individuals and thus equally increase the standard deviation.

The final hypothesis combines the two previous ones and states that the performance gap between men and women in situations that require creativity decreases in a competitive setting. The experimental results indicate that the initial gap in creativity, seen in the non-pressure exercise, does indeed decrease in the sense that men are able to catch up to the women in the number of original ideas that are generated in a competitive environment. But with women generating 3.0 valid answers on average compared to 4.8 answers by the men, an even bigger gap emerges. This performance difference could be explained partially by the nature of the experiment. Due to many factors, women are forced into a potentially uncomfortable situation. First and foremost, they need to individually compete against men. As stated earlier, women prefer piece rate incentives instead of a 'winner takes it all' approach (Niederle & Vesterlund, 2011).

Additionally, no teamwork was permitted, even though women generally prefer to work in groups (Gneezy et al., 2003). Especially single sex groups could possibly decrease the gap in creativity in these kinds of pressure situations (Gneezy et al., 2003) and may have led to results that provide more support for the presented hypothesis, by increasing female performance while not influencing men. The women also need to compete against other women which may leverage their potential. However, they still need to face male participants as well, thus preventing direct female competition.

The standard deviation for both women and men increased in the competitive environment, with the standard deviation of the women being even slightly higher. As can be seen in Figure 5-4, the group had several participants that underperformed by a great margin but also three participants who outperformed the average group. This means that while, on average, women perform worse than men, some excel and outperform the average male performance. This could indicate that some women still feel comfortable in competitive circumstances and are able to perform to their full potential.

7 Research Limitations and Implications

For this chapter, we first conducted a quasi-experiment, which means that participants knew that they were taking part in an experiment. Additionally, our sample size was relatively small; a larger sample size would lead to more reliable results. Consequently, our conclusion only has a small statistical relevance and could have been influenced by a large percentage of outliers. Our sample also consisted of students between the ages of 17 and 19 years. Hence, no conclusion can be drawn for different age groups. Furthermore, the students' social and geographic environment might be important for the outcome of the study and could not be taken into consideration.

Although it was advised that students do not talk about the experiment, they still had the possibility to share their impressions with other students during breaks, which could impact the results. Furthermore, although communication was prohibited during the experiment, to some extent it did still occur.

One of the most important limitations is that different participants had to be used in the different stages of the experiment. This means that we were not able to measure the creativity of the same student in both, the non-competitive and the competitive setting.

Further quantitative and qualitative research work should be carried out to confirm and expand the results obtained in this study. Ideally, this work would mostly consist of a larger sample size and differentiate between different social and age groups. The setting should be chosen in a way that prevents communication during the exercise and the same participants should take part in the exercise in both a non-competitive and a competitive setting. Besides confirming and detailing results on creativity of female and male participants in a competitive and non-competitive setting, we would recommend extending the study to different incentives (monetary with different values, non-monetary) to see if the type of incentive would directly influence the results.

Regarding implications on a broader scale, we can derive that on the one hand women perform the same in a non-competitive setting and on the other hand men increase their performance drastically in comparison to women in a mixed gender competition. The working world nowadays is characterized by a high degree of competition and thus men will feel more welcome in the current situation. Consequently, a change in thinking could contribute to a better integration of women in the work place. Competing in teams or single-gender competitions can already contribute to decreasing the gap. Furthermore, if a change is achieved, this could even contribute to questioning the effectiveness of the women's quota, which is often seen as too rigid or not the right solution in some situations.

References

Abraham, A. (2016). Gender and creativity: an overview of psychological and neuroscientific literature. *Brain Imaging and Behavior, 10*(2), 609–618.

Ackoff, R. L. and E. Vegara (1988). Creativity in Problem-solving and Planning. In R. L. Kuhn (Ed.), *Handbook for Creative and Innovative Managers* (pp. 77–89). McGraw Hill, New York.

Adams, K. (2005), The Sources of Innovation and Creativity, National Center on Education and the Economy.

Apicella, C. L., Demiral, E. E., & Mollerstrom, J. (2017). No Gender Difference in Willingness to Compete When Competing against Self. *American Economic Review, 107*(5), 136–140.

Babcock, L., & Laschever, S. (2003). *Women don't ask: Negotiation and the gender divide.* Princeton, N.J: Princeton University Press.

Baer, J., & Kaufman, J. C. (2008). Gender Differences in Creativity. *The Journal of Creative Behavior, 42*(2), 75–105.

Barber, B. M., & Odean, T. (2000). Trading is Hazardous to Your Wealth: The Common Stock Investment Performance of Individual Investors. *SSRN Electronic Journal.* Advance online publication.

Bertrand, M., Hallock, K. F., (2001): The Gender Gap in Top Corporate Jobs. *ILR Review*, 55, 3–21.

Beyer, S. (1990). Gender differences in the accuracy of self-evaluations of performance. *Journal of Personality and Social Psychology, 59*(5), 960–970.

Blau, F. D., & Kahn, L. M. (2000). Gender Differences in Pay. *The Journal of Economic Perspective*, *14*(4), 75–99.

Brophy, D. R. (2001). Comparing the Attributes, Activities, and Performance of Divergent, Convergent, and Combination Thinkers. *Creativity Research Journal*, *13*(3), 439–455.

Deci, E., Cascio, LV. F., & Krussel, J. (1973). Cognitive evaluation theory of some comments on the Calder and Staw critique. *Journal of Personality and Social* Psychology, *31*, 81–85.

Dimitriadis, E., Anastasiades, T., Karagiannidou, D., & Lagaki, Maria (2018). Creativity and Entrepreneuership: The role of Gender and Personality. *International Journal of Business and Economic Sciences Applied Research*. (11), 7–12.

Forisha, B. L. (1978). Creativity and imagery in men and women. *Perceptual and Motor Skills*, *47*(3 Pt 2), 1255–1264.

Gneezy, U., & Rustichini, A. (2004). Gender and Competition at a Young Age. *American Economic Review*, *94*(2), 377–381.

Gneezy, U., Niederle, M., & Rustichini, A. (2003). Performance in Competitive Environments: Gender Differences. *The Quarterly Journal of Economics*, *118*(3), 1049–1074.

Healy, A., & Pate, J. (2011). Can Teams Help to Close the Gender Competition Gap?: *The Economic Journal*, *121*(555), 1192–1204.

John, J. P. (2017). Gender differences and the effect of facing harder competition. *Journal of Economic Behavior & Organization*, *143*, 201–222.

Kim, K. H. (2006). Can We Trust Creativity Tests? A Review of the Torrance Tests of Creative Thinking (TTCT). *Creativity Research Journal*, *18*(1), 3–14.

Koestner, R., Zuckerman, M., & Koestner, J. (1989). Attributional Focus of Praise and Children's Intrinsic Motivation. *Personality and Social Psychology Bulletin*, *15*(1), 61–72.

Kosfeld, M., & Neckermann, S. (2011). Getting More Work for Nothing? Symbolic Awards and Worker Performance. *American Economic Journal: Microeconomics*, *3*(3), 86–99.

Lin, W.-L., Hsu, K.-Y., Chen, H.-C., & Wang, J.-W. (2012). The relations of gender and personality traits on different creativities: A dual-process theory account. *Psychology of Aesthetics, Creativity, and the Arts*, *6*(2), 112–123.

Niederle, M., & Vesterlund, L. (2007). Do Women Shy Away from Competition? Do Men Compete Too Much? *The Quarterly Journal of Economics*, *122*(3), 1067–1101.

Niederle, M., & Vesterlund, L. (2008). Gender Differences in Competition. *Negotiation Journal*, *24*(4), 447–463.

Niederle, M., & Vesterlund, L. (2011). Gender and Competition. *Annual Review of Economics*, *3*(1), 601–630.

Price, J. (2008). Gender Differences in the Response to Competition. *ILR Review*, *61*(3), 320–333.

Proctor, R. A. (1991). The Importance of Creativity in the Management Field. *British Journal of Management*, *2*(4), 223–230.

Runco, M. A., & Jaeger, G. J. (2012). The Standard Definition of Creativity. *Creativity Research Journal*, *24*(1), 92–96.

Sokolova, S. (2015). The Importance of Creativity and Innovation in Business. Retrieved February 9, 2018 from https://www.linkedin.com/pulse/importance-creativity-innovation-business-siyana-sokolova/

Julian Hertzler, Maximilian-Cyrus Mehrpour
4 Gender Differences in Intuitive Product Usage

1 Introduction

This chapter focuses on gender differences in technology management, in particular on the influence of gender roles on the acceptance and use of technical products. The motivation to take a closer look at this topic stems from the principle 'know thy user', which is of special importance, because different user groups, like men and women, work with information and make decisions about the use of technology in very different ways (Venkatesh, Morris, & Ackerman, 2000). In addition, the increasingly important topic of gender issues is to be examined more closely, in order to draw conclusions on possible influencing factors that affect the use of technology in the context of gender. In the past, science has often dealt with technology usage and gender differences. The results from previous research that are relevant to this chapter will be discussed in more detail in Section 4.2.

However, there is a gap between two major research streams in this context. On the one hand, research on the acceptance of the technology that is used in a product (Davis, 1989; Ajzen & Fishbein, 1980) is often needed in marketing or innovation management and is examined in more detail within the context of these disciplines to make special customer behavior visible. On the other hand, the issue of gender differences is being researched in sociology and psychology. Of course, there are also studies that combine both topics. However, we aim to address a gap in the extant literature by adding the factor of prior experience to the discussion of gender differences in technology acceptance.

This chapter is structured as follows. In the beginning, we present extant literature on the topic of gender differences. Subsequently, we try to explain the relationship between gender and the use of technological products, as well as prior experience and product use. Based on the findings from past studies, we then formulate our hypotheses, which are tested in a quasi-experiment. For this purpose, an office chair is used as the object of examination. The test subjects are tasked with finding and utilizing the functions of the office chair. There are some studies have already included an office chair in their research (Groenesteijn, Blok, Formanoy, de Korte & Vink, 2009; Hedge, 2016). After the results of the experiment are presented, the hypotheses are discussed, and possible influencing factors are explained. Finally, directions for further research are illustrated.

2 Literature Review

In the following section we are going to give insights into the most relevant topics by examining previous research. First, we discuss gender differences in general, before investigating the subject of gender and technology usage more in depth. Finally, we consider the topic of intuitive product use.

2.1 Gender differences

The general opinions about differences between men and woman occupy us every day. This includes the discussion about a quota for women in the workplace as well as prejudices that women do not perform as well as men in spatial thinking. A popular prejudice is furthermore that men understand technology faster and better than women. The topic of gender differences is widely discussed in society, but also in science. Researchers have been working on this topic for many years (Correll, 2002; Lehr, 2006). The whole topic is too extensive to be discussed completely within the scope of this chapter. Therefore, we highlight only the perspectives that are most important and show how gender differences or similarities possibly influence our experiment.

Interestingly, a number of researchers (Bem, 1974; Sieverding & Alfermann, 1992) support the fact that gender differences do not have such a big influence on our behavior, but rather on our personal characteristics. Gender is more of a role or stereotype in society. The literature helps us to solidify our point of view and our intentions. The Bem Sex Role Inventory model describes that there are male and female personality traits, but it is also possible that men and women have characteristics from both categories (Bem, 1974). This theory contradicts the classical theory, where men and women both naturally show a certain set of characteristics (Sieverding & Alfermann, 1992).

The results of Bem were also confirmed by Hyde (2005) who developed "the Gender Similarities Hypothesis". They state that there are not so many differences between men and women. According to their research, the issue of gender differences does not have as big of an influence on our behavior as society does, which is the factor that creates stereotypes around men and women (Hyde, 2005). This paper outlines gender similarities and stands in great conflict with the theory of gender differences, which states that men and women are psychologically very different. The gender similarity hypothesis describes that men and women are similar in most, but not in all psychological variables. Findings from meta-analyses in gender difference research support this statement (Hyde, 2005).

2.2 Gender and technology product use

Researchers have tried to make the acceptance of a product measurable and to identify possible influencing factors with different methods to determine why people decide to use a product or not.

Three models are to be mentioned here, that can be used in different ways. The Theory of Reasoned Action (TRA) (Ajzen & Fishbein, 1980) attempts to understand and predict human behavior. The Theory of Planed Behavior (TPB) (Ajzen, 1985) and the Technology Acceptance Model (TAM) (Davis, 1989) are derived from it. The Technology Acceptance Model and the Theory of Planed Behavior are the most important methods on which we base our research to understand the individual adoption and use of technology. However, the Technology Acceptance Model is the main method applied in our work. Davis (1989) uses the Technology Acceptance Model to examine software, but it can also be applied to measure technical products. This method is well suited as a basis for our experiment, since it considers many factors that also occur in our investigation. Technology acceptance describes in this context, whether a consumer uses a technology or not. The model is divided into two main components: The perceived usefulness and perceived ease of use. Perceived usefulness indicates whether the technology helps people to do their work. Ease of use describes whether the user perceives the technology as easy to use or too complex (Davis, 1989), which would naturally reduce the scope of use (Davis, 1989). However, the use of an object also depends on the attitude of the consumer (Fishbein & Ajzen, 1975). To understand the impact of gender differences in technology adoption and usage decisions, it is important to understand these underlying mechanisms first. Figure 4.1 depicts the Technology Acceptance Model.

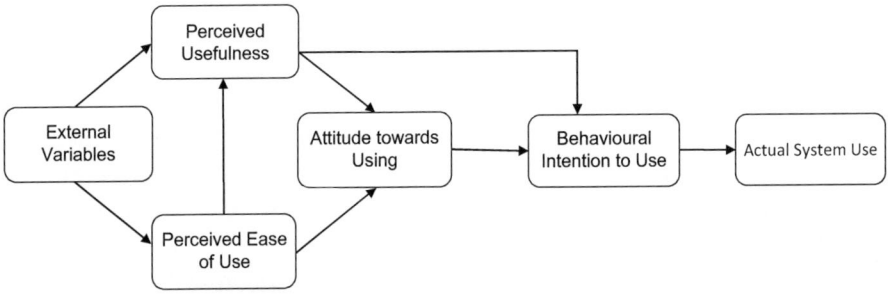

Figure 4.1: Technology Acceptance Model (Source: Own illustration based on Davis 1989).

Gender differences are not considered in the original Technology Acceptance Model, but there are modified versions that take this factor into account. Gefen and Straub (1997) tested the influence of gender differences on the technology acceptance of emails. The results show that sex has a measurable effect on the use of technology (Gefen & Straub, 1997). The factor gender has a robust influence on the components' perceived usefulness and perceived ease of use of the Technology Acceptance Model. Therefore, different technology acceptance models should take this factor into account (Venkatesh, Morris, Quarterly, Mar, & Morris, 2000).

Men and women also differ from each other in their risk-taking behavior. In this context, science is investigating the occurrence of psychological factors. In addition to features such as interest and curiosity, fearfulness is also mentioned as a factor (Kothgassner & Felnhofer, 2013). The reason to look at this aspect more closely is that there are large differences in the risk-taking behavior between the genders (Byrnes & Miller, 1999), which can also be applied to the factor fearfulness of women and men. It was found that women tend to aim for less risk in financial matters in order to have a higher level of security (Powell & Ansic, 1997). The results of an experiment by Byrnes and Miller (1999) underline the fact that males are more likely to take risks than females. The same experiment also showed that women and girls were rather unwilling to take risks even in harmless situations or when it was a good idea to take a little bit of risk. Thus, it can be deduced that the factor fearfulness shows significant differences between the two sexes because being risk-averse can be looked upon as someone being fearful of taking risks.

2.3 Intuitive product use

In order to understand the procedure followed in this work, the term intuitive product usage is explained in more detail in the following. There are many papers on the topic of the intuitive use of products, especially in the field of design thinking or universal design. Many researchers in this area use the word intuitive but seldom define it further. Just like in the work of Frank and Cuschieri (1997), which is about a gripper for surgeries, they describe that the new instrument handle was developed to work intuitively, but they do not define what intuitive means exactly. This is just one of many examples. The area of universal design also does not provide a useful definition. In the following work, the design of products is described, and seven principles are established. Principle three is 'Simple and Intuitive Use' (Molly Follette Story, Mueller, & Mace, 1998), but no further definition is given. Blacker provides a useful approach here, by saying that intuition is often subconsciously influenced and uses stored experience and knowledge; it fulfills the function of a cognitive processor (Blackler, Popovic, & Mahar, 2003). He refers back to Nardi (1996) who found that human experience is typically shaped by the tools and sign systems we use. As a conclusion, intuitiveness can be based on experiences that people have made with a product.

The intuitive use of a product by one person can be divided into three factors (Blackler et al., 2003). The first one is the location of the feature on the product. The second factor is the appearance of the product feature, which can include the shape, labeling, structure and colour. The third factor, the function of the feature refers to the way it works. Therefore, every factor must be considered for any feature of interest when it comes to the intuitive use of a product (Blackler et al., 2003). The intuitive use of a product is therefore defined in this work as the use of a product without prior instruction but with prior experience.

3 Hypotheses

The examined literature provides enough information to establish a comprehensive picture of the role of gender in the use of technologies and their acceptance. In the following, the hypotheses will be established that are relevant to this paper.

The first hypothesis that we want to test includes the preconceived notion that men are better at using technical products. We want to test whether there are possible gender differences. Therefore, we presume for the results of our experiment:

Hypothesis 1: Men will find more functions than women.

Examining the literature has shown that the impact of psychological factors such as fearfulness differs between the genders and that women are more risk-averse and thus in general more fearful than men. It can be assumed that this translates to all areas of life, which also includes the usage of products. Therefore, we believe that the risk-taking factor has a significant impact on the use of technology and that there are differences between men and women in this area. We are therefore testing this with our second hypothesis.

Hypothesis 2: Men are more likely to take risks concerning the usage of a product.

As it could be seen in the literature, previous experience plays an important role in the intuitive use of a product. Therefore, this factor needs to be considered as well in order to be able to make any statement about the way women and men use technology.

Hypothesis 3: The more prior experience a person has with a product, the more secure will be the use of a similar product.

In order to unite all these possible influences on the intuitive use of a technology, we test the developed hypotheses in a quasi-experiment.

4 Methodology

4.1 Experimental approach

For the experimental setup of our study, we decided to conduct an experiment, more specifically a quasi-experiment. It is a type of experiment, that does not allow for a randomization of the sample of participants.

Quasi-experiments are often useful for research that takes place in schools, because the class or the sample of pupils is already predefined before the experiment. On the one hand, quasi-experiments are convenient and practical, but on the other hand they reduce internal validity (Ross & Morrison, 2004).

4.2 Sample

Six boys and six girls from the Staatliche Fachoberschule und Berufsoberschule Erlangen, a specialized secondary school, were chosen as participants. The pupils were doing the experiment voluntarily and had no classes on that day. All the twelve participants had the specialization of social affairs and were between 17 and 19 years old, with an average of 17.7 years. The participants were not informed that we wanted to observe gender differences. They were told that the experiment was a part of a product prototype process to improve the office chair and to redevelop it, in order to evoke the feeling that the focus was on the product and not them.

4.3 Procedure

The study was conducted in a start-up office-environment at the Zollhof Tech Incubator in Nuremberg on the 26.04.2018 as part of the Girls' and Boys' Day. An office chair with eight different functions was chosen for the experiment (Figure 4.2). The challenge for the participants consisted of finding the eight different functions of the chair and to use them correctly, without any prior instructions or introduction.

In the experiment, we wanted to evaluate how prior experience may have influenced the product usage, since this is a factor that is important for the intuitive use of a product, according to the literature (Blackler et al., 2003). Additionally, we wanted to observe to what extent the participants where intuitive in the usage of the chair and how males and females differ in their approach to its usage.

The participants were tested separately, one at a time, in a room that entailed only the office chair. Three researchers were also present; two of them were executing the experiment with the participant while the remaining one was filming the whole procedure, for subsequent qualitative analysis. The procedure is described in the following paragraph.

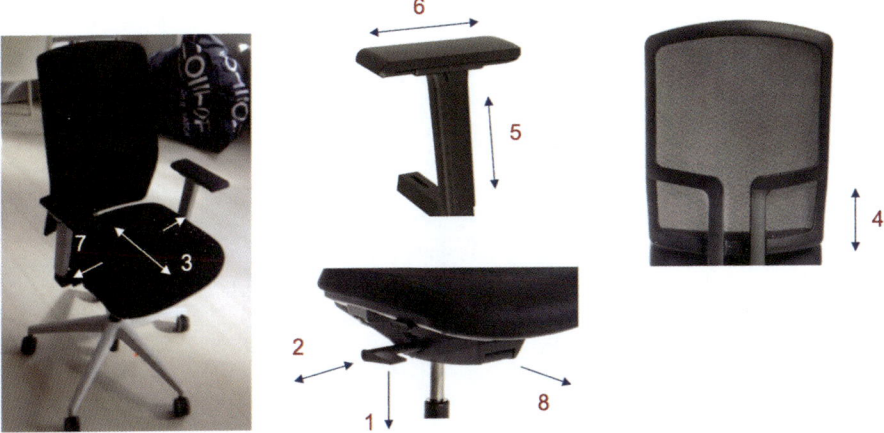

Figure 4.2: Chair used in the experiment with functions; 1 = seat height; 2 = backrest; 3 = seat depth; 4 = lumbar support; 5 = armrest height; 6 = armrest depth; 7 = seat width; 8 = backrest pressure (Source: Own illustration).

First, the participants were told that the experiment was a part of a product prototype process to improve the given product, i.e. the chair at hand. They were asked to fill out two question sheets; one before and one after the task. The questions were based on the components of the extended TAM, where the behavioral intention to use was tested (Gefen & Straub, 1997).

Additionally, to test if and to which extent the test subjects had prior experience concerning office chairs or chairs in general, the participants had to choose from three different kinds of chairs, depending on which one they were using most frequently in their daily life, e.g. at home or in school (see Figure 4.3). Each chair constitutes one level of complexity, from c1 (relatively easy to use) to c3 (more complex). The chair model c3 is representative of the chair, which has been used in the experiment from literature, which is why we also used a similar chair in our experiment.

The chair in our experiment was presented without giving information about it nor mentioning the amount of functions it had. The participants were told to interact with the chair by finding the different functions and using them correctly.

The time that the participants needed for this task was measured. After three minutes, the researchers were intervening, by listing the amount of functions that have been found and used correctly and the amount of functions that were still missing. After another three, and thus a total of six minutes, the experiment was stopped, followed by a more detailed explanation of all the functions.

In the third step, the participants were asked to repeat all the functions and focus on their perceived ease of use and usefulness. This information was recorded in the second question sheet. The perceived ease of use and perceived usefulness are

Figure 4.3: Frequently used chairs (Source: [Chair A59], n.d.; [LÅNGFJÄLL], n.d.; [Ikea Volmar + Armlehne], n.d.).

both components of the Technology Acceptance Model, which is why they were specifically included in the experiment.

After the experiment, the videos were qualitatively evaluated, to take a deeper look at how the participants were behaving and to see what the main challenges were and which gender differences concerning the finding and intuitive usage of the chair functions could be evaluated.

5 Results

This section shows the results of the experiment. This includes the functions that the participants found, the different components of the Technology Acceptance Model that could be observed in the experiment, as well as the results concerning prior experience with the chair that was used in the experiment.

5.1 Functions found by the participants

Table 4.1 is divided into boys (b) and girls (g) and it shows which participant found which function (f) of the office chair (f1 = seat height; f2 = backrest; f3 = seat depth; f4 = lumbar support; f5 = armrest height; f6 = armrest depth; f7 = seat width; f8 = backrest pressure); 'Sum' stands for the total amount of functions found by the participant. The sign 'x' in the table means that a function has been found and used correctly by the participant. A blank space indicates that the function has not been found and/or not used correctly.

Table 4.1: Functions found by the participants (Source: Own illustration).

	B1	B2	B3	B4	B5	B6	G1	G2	G3	G4	G5	G6
F1	x	x	x	x	x	x	x			x	x	
F2	x				x							
F3		x		x	x						x	x
F4	x	x		x	x			x	x		x	x
F5	x	x	x	x	x	x	x	x	x	x	x	x
F6	x	x	x	x			x	x		x		x
F7	x	x		x	x					x	x	x
F8		x										
Sum	6	6	4	6	6	2	3	3	3	4	5	4

As it can be seen in the table, the average amount of functions that were found amounts to 5 for the male participants and to 3.67 for the females. F2 and F8 were exclusively found and used correctly by boys. F5 was found by all the participants.

5.2 Technology Acceptance Model

Table 4.2 presents the different components of the Technology Acceptance Model, as well as the average answers given by the participating boys and girls.

Table 4.2: Components of the Technology Acceptance Model and the participants' answers (Source: Own illustration).

	Male participants	Female participants
Interest	4.0	3.7
Curiosity	2.6	2.8
Accessibility	4.6	4.3
Disbelief	1.7	2.1
Fearlessness	6.5	4.7
Ease of use	3.8	3.6
Usability	3.8	3.5

The questions of the survey were evaluated with a seven-point Likert scale ranging from a score of 1 (for strongly disagree) to 7 (for strongly agree). This scale is often used in questionnaires, since this specific number of possible answers prohibits a neutral response (Heijden, 2004).

As it can be seen, boys had a slightly higher interest to interact with the chair, felt like it was easier to use and rated the usability in general a bit higher. Girls had higher curiosity towards using the chair, but also more disbelief, which means that they were less confident in their ability to perform a certain action than the boys. The component that shows the biggest difference is fearlessness during the use of the chair.

5.3 Prior experience

In order to evaluate, which chair type the participants were most familiar with, the participants were given pictures of three chairs (Figure 4.3) with the task to mark the one they use most often in their daily lives (for school or office work) with an 'x'. They were only allowed to choose one of the options. The chairs are arranged according to their complexity, with c1 bein the least and c3 the most complex option. Table 4.3 shows the results of this questionnaire.

Table 4.3: Prior experience of the participants with the different chair types (Source: Own illustration).

	B1	B2	B3	B4	B5	B6	G1	G2	G3	G4	G5	G6
C1							x					x
C2						x		x		x		
C3	x	x	x	x	x				x		x	

The results reveal that five out of six boys use the chair type 3 most frequently to do their office or school work. The female participants had an equal preference for all three chair types with two girls per chair type. This means that the boys overall had more prior experience with chair type 3, which was the chair that was used in our experiment.

6 Discussion

In the literature, product usage has been a widely discussed term over the last centuries. The Technology Acceptance Model (TAM) by Davis (1989) is a fundamental model in this area, which describes that a system or product is used differently, depending on factors, such as the perceived usefulness and ease of use which forms the attitude towards using and thus leads to the actual use of a product. Additionally, in the literature there was a need to extend the original Technology Acceptance Model by the variable gender. Researchers point out that it does have an influence on the mentioned factors of the Technology Acceptance Model (Gefen & Straub, 1997).

As we can see from our results, the first hypothesis, which says that men will find more functions than women and use them correctly, could be supported. Within the scope of our experiment, it could be observed that boys and girls did have a different approach to the use of the product.

As it can be seen in Table 4.1, the seat height was found and used correctly by all boys, whereas just two out of six girls managed to do that. The seat height is usually a function that feels the most intuitive since it is one of the fundamental functions of office chairs. Looking at another study, where participants were given an office chair to see which of the adjustable function get used most often, the seat height was with a percentage of 92–95% the most used one (Groenesteijn et al., 2009).

During the qualitative examination of the video, it became evident that the girls had less problems with finding the function of the seat height and rather struggled with its usage. Together with other functions, like function 8 (backrest resistance), function 1 (seat height) was relatively hard to adjust, because more pressure than usual was needed to fully contract the lever. This was due to prototype reasons. This information was given to us by the company who made the chair. In contrast to the boys, who did not show much hesitation and tenderness when it came to the usage, the girls had obvious problems with the lever. They often said that they did not want to damage it. This leads to the conclusion that they were more fearful in the use of the lever of the chair. In contrast to that, function 5, the armrest height, was found and used correctly by all girls, since it was rather easily adjustable.

Hypothesis 2 that states that men are more likely to take risks than women concerning the usage of a product can thus be supported. As mentioned before in the literature review, men tend to take more risks than women and women sometimes do not take risks, even when the situation would be suitable for it (Byrnes & Miller, 1999). This is supported by the results in our study when finding the functions of a prototype chair.

Further support for Hypothesis 2 can also be found in Table 4.2. The results of the questionnaire of the extended Technology Acceptance Model that was filled out before the experiment, show that boys were more fearless compared to girls, concerning the use of the office chair.

Interestingly, the boys had more prior experience with office chairs in general (see Table 4.3). This could be one reason why the boys could find and use the functions of the chairs more confidently. Girls on the other hand had in general not as much experience with this type of office chair and found and used the functions less often and less correctly. This could suggest that past experience with the same product type leads to a better understanding of the use of a product of a similar type in the future without having read any kind of information about the product before (Blackler et al., 2003). So, our third hypothesis which states that a higher amount of prior experience leads to a more secure usage of a product can be supported.

During an experiment where the participants had to use a camera without prior instructions, the researchers concluded as well that participants with higher prior experience showed a more confident and better usage of the product (Blackler, Popovic & Mahar, 2003). Thus, the level of prior experience has a significant influence on the usage of the product.

7 Limitations and further research

One big limitation of the study is the sample size of the quasi-experiment. The number of pupils we conducted the experiment with, only amounted to twelve. Additionally, all the pupils were trained in one specific area of expertise. Thus, the experiment does not represent any kind of population group or randomly assigned participants. Furthermore, the participants had different levels of motivation regarding the experiment. Some of them gave up after one minute, so that the researchers had to intervene and ask them to continue. In the end, one must be cautious to transfer the gained knowledge in the study to the general population, as only a slight tendency could be observed. In order to really take more factors into consideration, it would be interesting to have more background information about the participants, such as experience with technical products or topics they are concerned with in their free time. Future research could increase the sample size of the study and use a chair model that requires less strength input, since this was one of the reasons that girls had problems with handling the office chair. Furthermore, we have tried to take as many factors as possible into account that affect the use of technology, but there will always be factors that cannot be taken into consideration within the scope of one experiment.

Further research could focus more intensively on one of the factors. The topic of gender differences is a very broad one, as it extends also into the cognitive behavioral research discipline. This research area has only been developed to a very small extent and offers potential for future investigations.

8 Conclusion

Ultimately it can be said that in the study a difference could be observed between boys and girls in the intuitive usage of the office chair. Boys did find more functions than girls, so that one could be tempted to conclude that they generally had more technical knowledge. However, considering the experience our participants had with the specific type of office chair that we used in our experiment, the gender gap becomes a smaller factor. The boys had more experience with office chairs that were like the one in the experiment, which is also a reason why they found and used more functions and thus had a better intuitive approach, but also showed a higher prior experience ranking. Furthermore, girls had a lower fearlessness ranking because they were more careful in the interaction with the chair, during the experiment. One plausible reason for that could be that they on the one hand had less prior experience and on the other hand were more afraid to damage it.

In the end it can be said, that a general gender difference in the usage of a product can be observed within the framework of this quasi-experiment. Girls had a different, more careful approach than boys. As mentioned before, this could have been due to a lack of prior experience among the girls. Thus, when considering all the results, it becomes evident that there may not be a large difference between the genders.

As described in the literature, factors like education, culture and social environment influence the behavior and characteristics of men and women. This gender gap most likely does not have a natural origin but stems more from the way we grow up and experience different kinds of products.

References

Ajzen, I. (1985). From Intentions to Actions: A Theory of Planned Behavior. In *Action Control* (pp. 11–39). Berlin, Heidelberg: Springer Berlin Heidelberg.

Ajzen, I., & Fishbein, M. (1980). *Understanding attitudes and predicting social behavior*. Prentice-Hall.

Bem, S. (1974). The Measurement of Psychological Androgyny. *Journal of Consulting and Clinical Psychology, 42*(2), 155–162.

Blackler, A., Popovic, V., & Mahar, D. (2003). The nature of intuitive use of products: an experimental approach. *Design Studies, 24*(6), 491–506.

Byrnes, J. P., & Miller, D. C. (1999). Gender Differences in Risk Taking: A Meta-Analysis.

[Chair A59]. (n. d.). Retrieved September 26, 2018 from https://thomasmoebel.eu/Chair-A59-p183

Correll, S. J. (2002). Gender and the Career Choice Process: The Role of Biased Self-Assessments. *American Journal of Sociology, 106*(6), 1691–1730.

Davis, F. D. (1989). Perceived Usefulness, Perceived Ease of Use, and User Acceptance of. *Information Technology MIS Quarterly, 13*(3), 319–340.

Fishbein, M. A., & Ajzen, I. (1975). Belief, attitude, intention and behaviour: An introduction to theory and research.

Gefen, D., & Straub, D. W. (1997). *Gender Differences in the Perception and Use of E-Mail: An Extension to the Technology. Source: MIS Quarterly* (Vol. 21). Retrieved from https://www.jstor.org/stable/pdf/249720.pdf?refreqid=excelsior%3A477042b27d89b959f157547c4ec87f50

Groenesteijn L., Blok M., Formanoy M., de Korte E., Vink P. (2009) Usage of Office Chair Adjustments and Controls by Workers Having Shared and Owned Work Spaces. In: Karsh BT. (eds) Ergonomics and Health Aspects of Work with Computers. EHAWC 2009. Lecture Notes in Computer Science, vol 5624. Springer, Berlin.

HeidelbergHedge, A. (2016). What Am I Sitting On? User Knowledge Of Their Chair Controls. *Proceedings of the Human Factors and Ergonomics Society Annual Meeting, 60*(1), 455–459.

Hyde, J. S. (2005). The gender similarities hypothesis. *American Psychologist, 60*(6), 581–592.

[Ikea Volmar + Armlehne]. (n. d.). Retrieved September 26, 2018 from https://www.testberichte.de/p/ikea-tests/volmar-armlehne-testbericht.html

Kothgassner, O. D., & Felnhofer, A. (2013). *Technology Usage Inventory*. Retrieved from https://www.ffg.at/sites/default/files/allgemeine_downloads/thematischeprogramme/programmdokumente/tui_manual.pdf

[LÅNGFJÄLL]. (n. d.). Retrieved September 26, 2018 from https://www.islas.ikea.es/ibiza/desktop/es_es/espacio-de-trabajo/sillas-de-trabajo/l%C3%85ngfj%C3%84ll/sillagiratoria/?&hfb=1004&ra=8035&item=156772&id_src=5

Lehr, J. L. (2006). *Athena Unbound: The Advancement of Women in Science and Technology. BioScience* (Vol. 51).

Molly Follette Story, Mueller, J. L., & Mace, R. L. (1998). *The Universal Design File Designing for People of All Ages and Abilities*. Retrieved from https://files.eric.ed.gov/fulltext/ED460554.pdf

Powell, M., & Ansic, D. (1997). Gender differences in risk behaviour in financial decision-making: An experimental analysis. *Journal of Economic Psychology, 18*(6), 605–628.

Ross, S. M., & Morrison, G. R. (2004). Experimental research methods. Handbook of research on educational communications and technology, 2, 1021–43.

Sieverding, Monika, & Alfermann, D. (1992). Geschlechtsrollen und Geschlechtsstereotype. *Zeitschrift Für Sozialpsychologie, 23*, 6–15.

Venkatesh, V., Morris, M. G., & Ackerman, P. L. (2000). A Longitudinal Field Investigation of Gender Differences in Individual Technology Adoption Decision-Making Processes. *Organizational Behavior and Human Decision Processes, 83*(1), 33–60.

Venkatesh, V., Morris, M. G., Quarterly, S. M. I. S., Mar, N., & Morris, M. G. (2000). Why Don't Men Ever Stop to Ask for Directions? Gender, Social Influence, and Their Role in Technology Acceptance and Usage Behavior. *MIS Quarterly, 24*(1), 115–139.

Sophia Ohmayer, Leonardo Reuther, Bangdi Wang

5 Spatial Reasoning Ability and Methodological Problem-Solving in STEM at an Intersection of Gender

1 Introduction

A gender imbalance in the workforce of the developed countries of the 1980s was often accredited to a lack of female education, besides the stereotypically assigned social roles. (Eagly & Steffen, 1984; Smith & Bachu, 1999). The sharp rise in educative reforms, compulsory primary and secondary school attendance, as well as accessible academic and higher education, have substantially eliminated a male dominance in education in developed countries. In 2016, female university graduates in the USA exceeded their male counterparts with a graduation rate of 56.8% to 43.2% (as cited in Statista, 2018). In Germany, the female to male ratio of university graduates accounts for a similar distribution of 51–49% (Statistisches Bundesamt, 2017).

However, the STEM (Science, Technology, Engineering and Mathematics) field is to this day mainly occupied by males. Men hold currently about 85% of all STEM-related jobs in Germany (Bundesagentur für Arbeit, 2018). The phenomenon of female underrepresentation in the field of STEM is described metaphorically by gender and education expert J. Blickenstaff to be a 'leaky pipeline'. The 'pipeline' illustrates the educational and vocational career path of men and women who have an interest in pursuing a STEM-related career. Some people 'leak out' at different stages of this pipeline, which means that they are not aiming for an occupation in this field anymore. For instance, this could be secondary school students, who expressed an interest in STEM subjects but might choose a completely different field of study in higher education. It is also possible that some university students in science programs may switch majors before graduation. It is paramount to note that women 'leak out' far more than men do (Blickenstaff, 2005).

This phenomenon is supported by the STEM graduation and employment ratio in Germany. Even though the female to male ratio of graduates amounts to 51–49% and the rate of employment of men and women is nearly equal with 46–54%, the female to male graduation ratio in STEM-related majors comes only to 28–72%. Male employment in STEM-related careers exceeds female employment by nearly six-fold with a female to male ratio of 15%–85% (Statistisches Bundesamt, 2017; Bundesagentur für Arbeit, 2018).

This chapter aims to further investigate the reasons for this imbalance between the genders in STEM-related careers. The structure is as follows: Section 5.2 depicts the current state of literature concerning the reasons for these differences between

https://doi.org/10.1515/9783110593952-005

the genders in STEM-related career paths. The hypotheses that this chapter will be evaluating are outlined in Section 5.3. In Section 5.4, the methodology concerning the experimental setup is illustrated and Section 5.5 presents the findings. Section 5.6 discusses these findings in the light of the extant literature, while a special focus will be on differences between males and females in spatial reasoning abilities and a methodological approach to problem-solving. Section 5.7 outlines the limitations of our experimental design, as well as implications for research and practice.

2 State of literature

2.1 Cognitive reasons for gender imbalances in the field of STEM

Blickenstaff (2005) remarks that women are not consciously filtered out of the STEM pipeline by people in a position of power. On the contrary, educational and occupational orientation are, by large, personal and individual choices. The foundation for career decisions is, to a large extent, achievement-related (Blickenstaff, 2005; Duncan et al., 2007; Eccles, 1994). Numerous modern, longitudinal studies have provided firm evidence that individual attitude, past achievements and academic performance are pronounced predictors of educational and occupational choices. (Else-Quest, Mineo, Higgins, 2013; Pekrun, Maier, Elliot, 2009). These separate factors do, however, exhibit a reciprocal effect. Positive emotions, such as joy and pride, which originate from successful achievements and academic performance, reinforce an affirmative attitude towards the respective field. The same, but inverse principle applies to negative emotions which individuals associate with their academic hurdles and shortcomings, such as, but not limited to anger, anxiety, hopelessness and shame. (Duncan et al., 2007; Pekrun et al., 2009; Daniels et al., 2009).

STEM-related occupations and academic success require a distinct cognitive skill set. Besides IQ, logic and reasoning, decisive predictors of success in STEM-related efforts include mathematical ability, spatial reasoning ability (including auditory and visual processing), as well as a methodological problem-solving approach. (Uttal & Cohen, 2012; Lefevre et al., 2010; Wilson & Shrock, 2001).

Numerous studies have been conducted to investigate the extent to which these cognitive abilities differ in women and men. The studies provide a range of biological (Coluccia & Luose, 2004; Bock & Kolakowski, 1973), and social reasons (Su, Rounds, Armstrong, 2009; Baenninger & Newcombe, 1995) to explain the differences in cognitive abilities as well as employing this reasoning to interpret the gender imbalance in STEM-related education and occupation.

For instance, Bock and Kolakowski (1973) provide evidence that the human spatial visualizing ability and the ability to manipulate visual images are influenced by an X-linked gene and testosterone levels. As X-linked mutations occur less frequently in women, it is reasoned that men develop a more pronounced spatial reasoning ability, especially with increased testosterone production with age.

In addition, Baenninger and Newcombe (1995) discuss differences in environmental inputs of males and females when growing up. Baenninger and Newcombe's meta-analysis supports a causal linkage between formal and informal experiences of male and female children and their mathematical and spatial abilities. This study classified the hobbies and activities of young adolescents as masculine sex-typed, feminine sex-typed and neutral. Examples for masculine sex-typed hobbies and activities were football, archery and building model airplanes. Feminine sex-typed hobbies and activities included ballet and knitting. Young adolescents that primarily practiced male-dominant hobbies scored better on both mathematical and spatial aptitude tests. Thus, Baenninger and Newcombe (1995) raised the argument that mathematical and spatial reasoning abilities can be built and improved by practicing hobbies that are classified as belonging to the masculine sex-type. To suggest that mathematical and spatial reasoning abilities are more prevalent in males from the beginning and that they are therefore more predominantly active in STEM-related occupations is thus not a valid assumption since spatial reasoning abilities can be actively trained.

The third vital predictor of success in STEM-related efforts, namely a methodological problem-solving approach, must also not be dismissed in the discussion of the gender imbalance problem in the field of STEM. Research regarding gender-specific differences in methodological problem-solving abilities shows different results than the study for mathematical and spatial reasoning abilities. A methodological problem-solving approach describes a 'compliance' with relevant, already established standards and norms (Venkatesh, Morris, Ackerman, 2000). Being congruent and adhering to existing scientific rules and norms is of crucial importance in STEM-related endeavors (Wilson & Shrock, 2001). Various empirical studies suggest that women display a greater process orientation while recognizing associations between different elements more easily (Sheppard, Hartwick, Warshaw, 1988; Rotter & Portugal, 1969). This is a preliminary indicator that women are better at solving a problem methodologically than men.

2.2 Spatial reasoning and methodological problem-solving within the context of gender

Potential for further research arises from the conflicting notion that men excel in spatial reasoning abilities, while women exhibit a more sophisticated methodological problem-solving approach. Both cognitive abilities are decisive predictors of success in STEM-related efforts. The scope of spatial reasoning and methodological problem-solving abilities will be detailed in the following. Thereupon, a research question will be laid out, that investigates the impact of gender in conditions where both cognitive skillsets are required.

The cognitive skillset of spatial reasoning has been a significant focal point of scientific and psychological research for nearly 100 years. However, there is no consensus on an accepted definition of spatial reasoning (Hegarty & Waller, 2005). Most research regarding spatial reasoning has adopted Carroll's (1993) definition, termed 'spatial visualization', which describes the human cognitive processes of "apprehending, encoding and mentally manipulating three-dimensional spatial forms" (Uttal & Cohen, 2012, p. 153; Carroll, 1993). Thus, spatial visualization is the ability to comprehend, recall or even generate two- and three-dimensional figures and mentally rotate and transform these objects. Vandenberg and Kuse's (1978) Mental Rotation Test (MRT) has been one of the most widely employed standardized testing methods to measure and compare spatial visualization abilities. The MRT consists of 24 two-dimensional drawings of three-dimensional geometrical objects, which are to be compared by test subjects (Vandenberg & Kuse, 1978).

Spatial visualization abilities are highly relevant in STEM-related disciplines. For instance, spatial skills correlate positively with good performance in the field of biology. Rochford (1985) found evidence to support the relationship between spatial abilities and underachievement in anatomy classes of second-year medical students. Students who struggled with visualizing, translating and rotating shapes would also score poorly on practical anatomy examinations (Rochford, 1985). Moreover, Hegarty (2007) confirms the importance of spatial thinking in medical performance and training. They attribute the success of laparoscopic surgery (i.e. minimally invasive surgery) to spatial visualization abilities and being able to comprehend medical images such as x-rays and magnetic resonance scans (Hegarty, 2007).

Carroll's (1993) definition and scope of 'spatial visualization' has been criticized by researchers, thus additions and amendments have been made, depending on the issue at hand. For instance, the term 'spatial perception' describes the necessary cognitive skillset in geology and geography. It explains the individual's capability to establish spatial relationships with one's environment. In this context, Orion, Benchaim and Kali (1997) investigated the performance of undergraduate students in earth science studies. Students who scored well on spatial perception tests were able to identify geological folds in mountains more easily and thus scored higher on final exams than their peers (Orion, Benchaim, Kali, 1997). Spatial perception tests include the Rod and Frame Test (RFT) or the navigation of a 'virtual maze' (Moffat, Hampson, Hatzipantelis, 1998).

Considering the dynamic nature of the definition of spatial reasoning, we want to extend it as well. The need to physically shape objects is of grave importance, especially in engineering, besides visualizing and mentally manipulating elements, e.g. in the case of rapid prototyping. This requires the ability to create, form and manipulate physical objects in relation to other objects or specifications by hand or with the assistance of tools. This ability borrows and adopts the necessary components of spatial visualization (transforming three-dimensional objects) and spatial perception (determine relationships between three-dimensional objects) and adds the extension of the motor skill. We believe that human motor control, and thus eye-hand coordination, should be included in the scope of spatial reasoning ability as well.

As stated earlier, a methodological problem-solving approach could be described as compliance with relevant norms that are important for the respective field. A method describes a procedure which is systematic and in alignment with established standards or principles. In a scientific problem-solving sense, a methodological approach is characterized by an adherence to upheld scientific norms, laws, principles and formulas. This includes, for example, the law of conservation of energy and charge in chemistry and physics, as well as the laws of gravitation and relativity. Solutions to scientific problems must align with established scientific norms, laws, principles and formulas. Not only does this require an individual to understand the relevant scientific basis he or she is working with, but also to identify and establish relationships between compatible scientific principles. A methodological problem-solving approach thus describes an individual's congruency with relevant scientific principles and the ability to formulate solutions within the boundaries of scientific laws (Venkatesh, Morris, Ackerman, 2000).

Brydges, Carnahan, Safir and Dubrowski (2009) assessed the technical skills of medical students in relation to their process orientation. One group of students was given the possibility to access self-guided instructional material in addition to their regular course work and material. The content of the self-guided instructional material focused on medical process goals. Students who viewed the self-guided instructional material acquired psychomotor skills more easily and performed far better in transfer tests than their peers who did not view the self-guided instruction material. Brydges et al. (2009) thus support the theory, that comprehending process-related goals, especially gained in a self-guided, independent way, correlates positively with applying clinical technical skills.

In this context, Pauli, Drollinger-Vetter, Hugener and Lipowsky (2008) found that a methodological problem-solving approach contributes to learning success in mathematics. They primarily intended to analyze differences in teacher-guided vs. discovery-oriented instructions of 8th/9th-grade students from German and Swiss schools. They established that instructional quality does not depend upon the method of conceptual construction. However, through this experiment, it could be discovered that there is a relation between instructional orientation and learning success in mathematics. Students who displayed a clear understanding of the conveyed instructions also scored higher on mathematical examinations. Furthermore, students, who had understood the instructions, stated that learning and understanding new mathematical concepts was easier (Pauli et al., 2008).

3 Research question and hypotheses

Considering the previously defined factors of spatial reasoning and methodological problem-solving, we sought to investigate how gender affects these two factors, in a STEM-related context. In the following, we intend to deduce hypotheses from the extant literature concerning the underrepresentation of women in STEM-related occupations. Thus, we established the following research question:

To what extent does gender implicate methodological problem-solving in a STEM-related spatial reasoning context?

Based on the examination of extant literature, the hypothesis can be derived that men will outperform women in a STEM-related problem-solving task since the more pronounced spatial reasoning abilities in males are more pivotal (Rochford, 1985; Hegarty, 2007). However, women will be better at approaching the problem methodically than men.

4 Methodology

4.1 Participants

Overall, there were ten participants in our study. They were all students of a German vocational secondary school from the ages of 16 to 21 years (average age: 17.3 years). They were evenly divided by gender, so there was one group with five female and a second group with five male test subjects. Due to the nature of the experiment, personal profiles and past experiences may affect the experimental results. Therefore, a questionnaire was handed out prior to the experiment. The vocational school has two branches of education, namely economics and social sciences. 80% of the male and 40% of the female students had an economic background. Consequently, 20% of the male students and 60% of the female students studied social sciences. None of the participants had primarily STEM-related classes.

Additionally, previous experience with manual labor and craftsmanship may have an influence on the results of the experiment, since interaction with different materials was required. None of the male test subjects associated themselves with a craftsmanship background. However, 60% of the female test subjects stated that they had substantial experience with craftsmanship and considered it a hobby and recreational activity. As forming clay is a primary component of the experiment, it was of high interest to investigate whether the participants had worked with forming clay before. 40% of the male test subjects and 60% of the females answered this question positively.

4.2 Procedure

The test subjects had to solve a scientific problem. They were tasked with building a buoyant device that can carry a dummy object above water. A standardized amount of forming clay served as the basis for the buoyant device. A choice of additional materials was offered, which the students could integrate into their buoyant device. The dummy object was also standardized in shape and mass.

The experiment began by seating the test subjects separately, as the task was designed to be solved individually. The students were given the forming clay and the goal of the task was explained as follows:

> Your goal is to build a buoyant device with the forming clay that can carry the dummy object above water. The entire forming clay must be used. With the provided container of water, the buoyancy can be checked as often as desired, during the entire process. After the instructions, you will have the possibility to choose two additional materials from a set of ten different materials that you may integrate into the device. The use of these *assisting materials is not mandatory*. You will have one minute to choose any two assisting materials and another ten minutes to build the buoyant device.

The ten different assisting materials were hidden until after the test subjects had received the instructions. The assisting materials were classified in two categories depending on their properties. Five objects had buoyant characteristics. This includes balloons, styrofoam-balls, straws, tinfoil and wood sticks. The other five objects had a connective function: nails, elastic bands, cable tie, toothpicks and adhesive tape. It is crucial to note that these classifications and properties of the assisting materials were not revealed to the test subjects. Furthermore, the materials were presented mixed and randomly.

The subjects were asked multiple questions to demonstrate their approach and to give feedback on whether they had understood the scientific principles and relationships. After the test subjects selected their assisting materials, we inquired: "Why did you choose the aiding materials?"; "Are you content about your decision regarding the aiding materials?" After successful or unsuccessful completion of the experiment, the students were given the possibility to reflect upon their work and display their understanding of the problem. The following questions were asked: "Are you content with your solution?"; "What would you have done differently?"; "Would you have used different assisting materials?"

4.3 Measures

We applied four different measures during and after the experiment for each test subject. Three of the measures were evaluated binarily and the fourth measure is a time and duration-based measure. A binary evaluation describes a result which can only consist of two outcomes.

The first measure was the successful solving of the posed task. The solution was considered successful if the built buoyant device could carry the dummy object above water. The choice of assisting materials was not of relevance here. This measure was designed to reflect the spatial reasoning ability of the test subject, as a successful solution demonstrates the ability to transform and manipulate a three-dimensional object mentally and physically. This measure was recorded with either a 'Yes' or a 'No'.

The second measure, the ideal choice of assisting materials was the second binary measure. It describes whether the test subject was able to discern between the two different groups and identify the buoyant and connecting property of each material. This measure was designed to reflect the methodological problem-solving ability of each test subject, as an ideal choice of assisting materials demonstrates that participants understood the scientific properties of the materials and their relationship with each other. Whether their device achieved buoyancy or not, has no influence on the outcome of this measure. This measure was recorded with either a 'Yes' or a 'No'.

A valid reflective self-assessment was the third binary measure. It is a summary of all the questions which were asked and intends to support the previous measures. The test subjects were asked to explain, whether their choice of materials was scientifically reasoned or done arbitrarily. Furthermore, the questions intended to investigate whether the test subjects had really understood the problem and the relationship between the forming clay, the assisting materials and the water. This measure was recorded with either a 'Yes' or a 'No'.

The time to solve the posed problem is the only measure based on an SI-unit. It records the time each test subject needed to complete the experiment in seconds. The duration of a task is a valid indicator for effectiveness and efficiency.

4.4 Reasoning to design an interactive, multimodal experiment

The reason, why the experiment was set up in this specific interactive and multimodal way, was to promote genuine test subject performance and thus attain more authentic results.

Johns, Schmaders and Martens (2005) provided empirical evidence that informing women about stereotypes that convey a negative picture of women's mathematical competencies, has an adverse effect on their subsequent performance. In their experiment, female and male test subjects had to complete mathematical problems that were either labeled a problem-solving task or a math test. A third group was also integrated that was formed from a part of the female test subjects. Besides having the task described as a math test, they were also informed about the stereotype that women had a lower mathematical performance than men in general. Female test scores did not differ from male test scores when the task was described as a problem-solving task. However, women scored lower when it was described as a mathematical test and even lower when negative gender stereotypes were mentioned. Johns, Schmaders and Martens (2005) attributed female anxiety and the resulting lower performance to gender stereotypes that circulate in the STEM field. They also offer suggestions on how experiments investigating gender differences should be designed (Johns, Schmaders, Martens, 2005).

On this basis, we intended to design a more abstract experiment to distance the experimental instructions and execution from gender stereotypes that may arise, which could cause less representative results. We intended to focus the measures on spatial reasoning ability and methodological problem-solving. By making the experiment interactive and multimodal, we hoped for more motivation among the test subjects and consequently a personal impulse to accomplish the task to the best of their abilities.

5 Results

60% of the male test subjects and 40% of the female test subjects finished the task successfully (see Table 5.1). As mentioned before, this measure evaluates the spatial reasoning abilities of the test subjects. The ideal choice of assisting material was made by 40% of the male subjects and by 80% of the female subjects. The questions intending to investigate whether this choice was based on scientific reasoning or chance were posed subsequently. This measures the valid reflective self-assessment figure, which 40% of male test subjects achieved, as well as 60% of the female participants. The ideal choice of assisting materials and a valid reflective self-assessment are indicators of a methodological-problem solving approach. The average time to solve the problem hints at efficiency and effectiveness. Female test subjects were, on average, 15% faster than their male counterparts.

Table 5.1: Results from the experiment (Source: Own illustration).

	Male participant (n = 5)	Female participant (n = 5)
Problem successfully solved	60%	40%
Ideal choice of assisting materials	40%	80%
Valid reflective self-assessment	40%	60%
Average time to solve problem (in seconds)	526.6	445

6 Discussion

Even though women excelled in three of the four measurement categories, as they were able to choose ideal assisting material groups and establish their understanding of scientific relationships, men outperformed women in successfully solving the given task. The collected data offers support for our previously established hypothesis that men will outperform women in a STEM-related methodological problem-solving task, as the more pronounced spatial reasoning abilities in males are more pivotal than a methodological approach.

In order to find factors that can explain the results in view of the presence of gender differences in STEM-related education and vocation, we considered both spatial reasoning abilities and a methodological problem-solving approach.

6.1 Gender differences concerning spatial reasoning ability

The more distinct spatial reasoning abilities in males, which can also be seen from the results, can be explained in multiple ways. Three major reasonings will be discussed in the following. These include a biological aspect, a social aspect and an interactionist theory.

As mentioned in the beginning, there is substantial evidence from medicine that the visualization of spatial relations is influenced positively by an X-linked gene and levels of testosterone. Bock and Kolakowski (1973) further investigated the effect of X-linked genes based on previous studies, such as those done by Kelley in 1927. On the basis that spatial abilities are different from verbal abilities and general intelligence, they found empirical evidence that suggests that X-linked traits include more distinct spatial abilities. As mutations in the respective genes occur less frequent in women, it is reasoned that men develop a more pronounced spatial reasoning ability (Bock & Kolakowski, 1973). Sexual hormones and their level also have an influence on the cognitive development of individuals. For instance, Moffat, Hampson and Hatzinpatelis (1998) conducted experiments, in which male and female participants were asked to navigate a 'virtual maze'. Their performance, measured by the time they needed to complete the task, was compared to the testosterone and estrogen levels of the subjects. Generally, male participants outdid females. Additionally, males who scored in the top 15-percentile also had the highest testosterone levels. Female ovaries do produce testosterone but in seven to eight times lower amounts. Moffat, Hampson, Hatzipantelis (1998) concluded on a causal linkage between spatial abilities and individual testosterone levels. A third argument from the area of medicine for more pronounced spatial reasoning abilities in men is accredited to certain cerebral patterns in the left hemisphere of the brain. These positively foster language skills but have an adverse impact on the development of spatial awareness skills. Girls at a young age develop the left hemisphere earlier, which could result in less developed spatial reasoning abilities, but have a positive impact on language skills (Annett, 1992).

Nevertheless, environmental inputs from society, culture and race have a strong influence on the spatial abilities of males and females. Baenninger and Newcombe raised the argument that mathematical and spatial reasoning ability can be trained. This training often occurs subconsciously, e.g. through the practice of hobbies. Baenninger and Newcombe (1995), as well as other scholars such as Lawton and Morrin (1999), suggest that 'masculine' hobbies such as football, archery or LEGO-construction stimulate the spatial reasoning abilities of individuals. Thus, mathematical and spatial reasoning abilities could be built and improved by practicing said activities. As men generally have more experience in activities that enhance the development of spatial skills, men outperform women, due to more time spent doing these activities (Baenninger & Newcombe, 1995; Lawton & Morrin, 1999). Contemporary research also suggests that playing three-dimensional action video games can reduce gender differences in spatial cognition, further consolidating the claim that spatial abilities can be trained and are influenced by hobbies and activities (Feng, Spence, Pratt, 2007).

The interactionist theory provides the third argument, which asserts that the reason for gender differences is the continuous interaction between biological and environmental factors. Since males are naturally more interested in spatial activities, they also invest more time in them. The execution of these activities consequently leads to an increase in the initial difference in spatial reasoning abilities (Coluccia & Louse, 2004).

6.2 Gender differences concerning methodological problem-solving

The choice of ideal assisting materials and a valid reflective self-assessment are indicators of a methodological problem-solving approach. As defined in the introduction, a methodological problem-solving approach describes an individual procedure, which is systematic and in alignment with established standards or principles, to solve a given problem. In a scientific sense, it is characterized by an adherence to upheld scientific norms, laws, principles and formulas. In our experimental design, this parameter was measured through the correct identification and implementation of the buoyant and connecting properties of the assisting materials, as well as the recognition of the physical relationship between the object's shape and its capabilities of floating above the water. Various empirical evidence suggests that women display a greater process orientation, while recognizing associations between elements more easily, agree with the data we collected (Sheppard, Hartwick, Warshaw, 1988; Rotter & Portugal, 1969). Women also needed less time to solve the given problem. This is an indicator that a methodological problem-solving approach is beneficial for the effectiveness and efficiency of the process.

Differences in gender regarding methodological problem solving can also be attributed to biological and social reasons.

Speck, Ernst, Braun, Koch, Miller & Chang (2000) investigated gender differences from a biological perspective by analyzing the functional organization of the brain for working memory. Working memory is the capability of the human brain to assess and manipulate information from different sources and draw relationships between them. Working memory "is an important component for many higher cognitive functions" (Speck et al., 2000, p. 2581). The volume of activated brain tissue was measured while test subjects were given increasingly difficult verbal tasks. Male participants showed bilateral activation or dominance from the right-sided brain hemisphere. Females showed activation predominantly in the left hemisphere. The results showed that women demonstrated more accurate solutions to the given problems. The left-brain hemisphere was accredited to be more influential regarding working memory and the development of problem-solving strategies (Speck et al., 2000).

Differences in methodological problem solving can also be interpreted from a social point of view. Barrick and Mount's (1991) widely acclaimed 'Big Five Personality Dimensions' (extraversion, emotional stability, agreeableness, conscientiousness and openness to experience) serves as a basis for this argument. The personality dimensions of agreeableness and openness to experience correlate with a methodological problem-solving approach. Empirical evidence shows a general trend towards being agreeable and away from openness to experience in women (Barrick & Mount, 1991). Women, when confronted with problems, will associate a solution with past experiences. On the other hand, "men are more likely to be willing to put more effort to overcome constraints in order to achieve their objectives without necessarily thinking about or emphasizing the magnitude of the effort involved" (Venkatesh, Morris, Ackerman, 2000, p. 38). Gustafson (1998) supports this statement in his analysis of gender differences in risk perception. In his studies, he finds support that men perceive risks far less frequently and are generally willing to take more risk than women (Gustafson, 1998).

7 Limitations and implications

7.1 Shortcomings regarding test subjects

Even though the quoted literature corresponds with the results of our quasi-experiment, some limitations need to be addressed. The limited number of test subjects (n = 10) is not representative to make conclusive statements about gender differences, especially not for all STEM applications. Furthermore, none of the participants had a STEM-related educational background, as the test subjects were comprised of students of a vocational secondary school with either an economic or social science area of study. The age range of just four years additionally limits the expressiveness of the experiment. In order to make this study more representative, a higher number of test subjects with different academic backgrounds in a larger age range would offer more conclusive results.

7.2 Shortcomings regarding the experimental design

Since a certain target population is part of a quasi-experiment, no random assignment could take place. This deficiency in randomization increases the difficulty to rule out confounding variables and introduces new threats to internal validity. Furthermore, causal relationships are difficult to determine. It would be necessary to achieve a clearer understanding of the direction of causation to draw a firm conclusion. Furthermore, spatial reasoning ability and a methodological problem-solving approach could only partly be measured isolated from other influencing factors. For instance, the measure of successfully solving the task is not solely dependent on spatial reasoning. Since a variety of factors and environmental contexts contributes to the emergence of different results, a causal interpretation of the findings is difficult.

During experimentation, we noticed that the provided assisting materials were not exclusively used in the way we intended to. For instance, one participant used tin foil as a connecting material, even though it was supposed to be used as a buoyant material. Thus, an improvement could be a clearer distinction between the properties of the assisting materials. Furthermore, the instructions were given by different people, which may lead to unwanted divergences, as the instructions might slightly differ in content and clarity. This can be improved by handing over the instructions in a written form or through a digital medium such as pre-recorded videos.

7.3 Further research: Importance of female-male balance in STEM workplaces

Provided that the underrepresentation of women in STEM disciplines persists and that team collaboration in scientific innovations are getting increasingly important, it is strongly suggested that an improvement of team collaboration is achieved by including women in the group (Bear & Woolley, 2013). A study investigating group performance in a business simulation found that groups with an equal number of men and women, as well as groups with a higher number of women than men performed better than homogeneous groups. This result was explained by more effective collaborative group processes and cooperative norms (Fenwick, Graham & Neal, 2001).

Furthermore, in contrast to methodological problem-solving, spatial ability is a personal trait that can be improved by using several tools and methodologies, like virtual and augmented reality or three-dimensional video games (Feng, Spence, Pratt, 2007). Due to the significance of this skill, it is crucial to foster and further develop the spatial reasoning skills in males and females alike. As spatial abilities are a determining factor in STEM-related vocation, private and public education should aim to stimulate the growth of spatial reasoning abilities and consequently hope to 'reduce the leakage' along the STEM-pipeline.

References

Annett, M. (1992). Spatial ability in subgroups of left-and right-handers. *British Journal of Psychology, 83*(4), 493–515.

Baenninger, M., & Newcombe, N. (1995). The role of experience in spatial test performance: A meta-analysis. *Sex Roles, 20*(5-6), 327–344.

Barrick, M. R., & Mount, M. K. (1991). The Big Five Personality Dimensions and Job Performance: A Meta-Analysis. *Personnel Psychology, 44*(1), 1–26.

Bear, J. B., & Woolley, A. W. (2013). The role of gender in team collaboration and performance. *Interdisciplinary Science Reviews, 36*(2), 146–153.

Bock, R. D., & Kolakowski, D. (1973). Further evidence of sex-linked major-gene influence on human spatial visualizing ability. *American Journal of Human Genetics, 25*(1), 1–14.

Brydges, R., Carnahan, H., Safir, O., & Dubrowski, A. (2009). How effective is self-guided learning of clinical technical skills? It&s all about process. *Medical Education, 43*(6), 507–515.

Bundesagentur für Arbeit. (2018). Der Arbeitsmarkt in Deutschland - MINT-Berufe. *Statistik der Bundesagentur für Arbeit*.

Carroll, J. B. (1993). *Human Cognitive Abilities: A Survey of Factor-Analytic Studies*. Cambridge: Cambridge University Press.

Clark Blickenstaff, J. (2005). Women and science careers: Leaky pipeline or gender filter? *Gender and Education, 17*(4), 369–386.

Coluccia, E., & Louse, G. (2004). Gender differences in spatial orientation: A review. *Journal of Environmental Psychology, 24*(3), 329–340.

Daniels, L. M., Stupnisky, R. H., Pekrun, R., Haynes, T. L., Perry, R. P., & Newall, N. E. (2009). A longitudinal analysis of achievement goals: From affective antecedents to emotional effects and achievement outcomes. *Journal of Educational Psychology, 101*(4), 948–963.

Duncan, G. J., Dowsett, C. J., Claessens, A., Magnuson, K., Huston, A. C., Klebanov, P., Japel, C. (2007). School readiness and later achievement. *Developmental Psychology, 43*(6), 1428–1446.

Eagly, A. H., & Steffen, V. J. (1984). Gender stereotypes stem from the distribution of women and men into social roles. *Journal of Personality and Social Psychology, 46*(4), 735–754.

Eccles, J. S. (1994). Understanding Women&s Educational and Occupational Choices: Applying the Eccles et al. Model of Achievement-Related Choices. *Psychology of Women Quarterly, 18*(4), 585–609.

Else-Quest, N. M., Mineo, C. C., & Higgins, A. (2013). Math and Science Attitudes and Achievement at the Intersection of Gender and Ethnicity. *Psychology of Women Quarterly*, *37*(3), 293–309.

Feng, J., Spence, I., & Pratt, J. (2007). Playing an action video game reduces gender differences in spatial cognition. *Psychological Science*, *18(10)*, 850–855.

Fenwick, Graham, D., & Neal, D. 2001. Effect of gender composition on group performance. *Gender, Work and Organization*, *8 (2)*,205–225.

Gustafson, P. E. (1998). Gender Differences in Risk Perception: Theoretical and Methodological perspectives. *Risk Analysis*, *18*(6), 805–811.

Hegarty, M & Keehner, M., Cohen, C. A., Montello, D.R. & Lippa, Y. (2007). The Role of Spatial Cognition in Medicine: Applications for Selecting and Training Professionals. Book chapter: Applied Spatial Cognition. 285–315.

Hegarty, M., & Waller, D. A. (2005). Individual Differences in Spatial Abilities. In P. Shah & A. Miyake (Eds.), *The Cambridge Handbook of Visuospatial Thinking* (pp. 121–169). Cambridge: Cambridge University Press.

Johns, M., Schmader, T., & Martens, A. (2005). Knowing is half the battle: Teaching stereotype threat as a means of improving women&s math performance. *Psychological Science*, *16*(3), 175–179.

Kelley, T.L. (1927). Interpretation of educational measurements. New York: World Book Company.

Lawton, C. A., & Morrin, K. A. (1999). Gender Differences in Pointing Accuracy in Computer-Simulated 3D Mazes. *Sex Roles*, *40*(1/2), 73–92.

Lefevre, J.-A., Fast, L., Skwarchuk, S.-L., Smith-Chant, B. L., Bisanz, J., Kamawar, D., & Penner-Wilger, M. (2010). Pathways to mathematics: Longitudinal predictors of performance. *Child Development*, *81*(6), 1753–1767.

Moffat, S. D., Hampson, E., & Hatzipantelis, M. (1998). Navigation in a "Virtual" Maze: Sex Differences and Correlation with Psychometric Measures of Spatial Ability in Humans. *Evolution and Human Behavior*, *19*(2), 73–87.

Orion, N., Benchaim, D., & Kali, Y. (1997). Relationship Between Earth-Science Education and Spatial Visualization. *Journal of Geoscience Education*, *45*(2), 129–132.

Pauli, C., Drollinger-Vetter, B., Hugener, I., & Lipowsky, F. (2008). Kognitive Aktivierung im Mathematikunterricht. *Zeitschrift Für Pädagogische Psychologie*, *22*(2), 127–133.

Pekrun, R., Elliot, A. J., & Maier, M. A. (2009). Achievement goals and achievement emotions: Testing a model of their joint relations with academic performance. *Journal of Educational Psychology*, *101*(1), 115–135.

Rochford, K. (1985). Spatial learning disabilities and underachievement among university anatomy students. *Medical Education*, *19*(1), 13–26.

Rotter, G. S., & Portugal, S. M. (1969). Group and individual effects in problem solving. *Journal of Applied Psychology*, *53*(4), 338–341.

Sheppard, B. H., Hartwick, J., & Warshaw, P. R. (1988). The Theory of Reasoned Action: A Meta-Analysis of Past Research with Recommendations for Modifications and Future Research. *Journal of Consumer Research*, *15*(3), 325.

Smith E., & Bachu A. (1999). Women&s Labor Force Attachment Patterns and Maternity Leave: A Review of the Literature. Washington DC: U.S. Bureau of the Census.

Speck, O., Ernst, T., Braun, J., Koch, C., Miller, E., & Chang, L. (2000). Gender differences in the functional organization of the brain for working memory. *Neuroreport*, *11*(11), 2581–2585.

Statistisches Bundesamt. (2017). Statistisches Jahrbuch. Retrieved 8 March 2018 from https://www.destatis.de/DE/Publikationen/StatistischesJahrbuch/Arbeitsmarkt.pdf?__blob=publicationFile

Statista. (2018). Educational attainment distribution in the United States from 1960 to 2017. Retrieved February 9 from https://www.statista.com/statistics/184260/educational-attainment-in-the-us/

Su, R., Rounds, J., Armstrong, P. I. (2009). Men and Things, Women and People: A Meta-Analysis of Sex Differences in Interests. *Psychological Bulletin 135(6)*, 859–884.

Uttal, D. & Cohen, C. (2012). Spatial Thinking and STEM Education: When, Why, and How?. *Psychology of Learning and Motivation*, 148–181.

Vandenberg, S. G., & Kuse, A. R. (1978). Mental rotations, a group test of three-dimensional spatial visualization. *Perceptual and Motor Skills, 47*(2), 599–604.

Venkatesh, V., Morris, M., Ackerman, P. (2000). A Longitudinal Field Investigation of Gender Differences in Individual Technology Adoption Decision-Making Processes. *Organizational Behavior and Human Decision Processes, 83(1)*, 33–60.

Wilson, B. C., & Shrock, S. (2001). Contributing to success in an introductory computer science course. In H. Walker, R. McCauley, J. Gersting, & I. Russell (Eds.), *Proceedings of the thirty-second SIGCSE technical symposium on Computer Science Education - SIGCSE &01* (pp. 184–188). New York, USA: ACM Press.

Philipp Röll, Michael Heimes
6 Gender Differences in Approaching and Solving Technical Tasks – An Experimental Research

1 Introduction

The STEM-disciplines (science, technology, engineering and mathematics) are widely seen as a fundamental factor to ensure the future innovative strength and global competitiveness of Germany. However, a significant shortage of skilled professionals has been occurring especially in this field for years. In September 2017, a new all-time high of 290,900 unoccupied positions in Germany was reached. 34% of these positions were academic ones. The field of engineering takes the highest share of this percentage, even though the number of graduates in the field of engineering highly increased in recent years. In this regard, it has to be emphasized that women are still highly underrepresented, especially in the field of engineering studies. The number of female graduates stagnated to a level of only 22% in the field of engineering. Accordingly, the percentage of women who graduated in the whole field of STEM-related studies only accounted for 29.7% in 2016 (Anger, Berger, Koppel, & Plünnecke, 2017).

In order to counter this development, it should be further investigated, how to increase the number of women working in STEM disciplines. Therefore, the chair of Technology Management at the Friedrich Alexander University of Erlangen-Nuremberg created the 'International Technology Management Research Seminar' with the topic of gender differences in innovation, in the summer term of 2018. The aim was to investigate gender differences in the fields of creativity, technical problem solving and technology acceptance and how to support women in the field of STEM. For this purpose, six teams with students from different academic backgrounds implemented and conduced experimental research projects in cooperation with three companies, namely Bock 1, Staedtler and UVEX. Within this framework, the pupils of the FOS/BOS Erlangen were invited as research participants to the "International Girls' and Boys' Day" at Zollhof in Nuremberg, a tech incubator. This event was organized by the chair of Technology Management with support and directions from the research associates.

This chapter investigates gender differences in approaching and solving technical tasks. It is divided into seven sections. First of all, a literature review is provided to determine the current state of research. After that, the research question is presented as well as three hypotheses for the outcome of the experiments. In the third section, the methodology of the three conducted experiments is explained. This is followed by the presentation of the results. Consequently, the results will be discussed. In the final section, a critical review and an outlook for future research will be given.

https://doi.org/10.1515/9783110593952-006

2 State of literature

In order to analyze an individual's behavior towards technical problem solving, the term 'problem' must be defined first. According to Savransky (2000) and Dörner (1984), a problem describes the gap between the initial state and the goal situation. Problem-solving is the transformation process between both states. This transformation process is executed in a single or multiple steps. If all the steps which are necessary to accomplish the task are known, the process can be labeled a routine. If there is at least one unknown step, the individual is forced to develop own problem-solving strategies. Other explanations and definitions of problem-solving, especially technical problem solving, also consider the impact of uncertainty and ambiguity (Schrader, Riggs, & Smith, 1993). In this context, uncertainty describes multiple possible results, which can be ranked subjectively, while ambiguity refers to a lack of clarity (Duncan, 1972; Marples, 1961).

According to Miller, Kelly, and Kelly (1988) there is a strong correlation between technical problem solving and the ability to discern between spatial relations. This is best described with being capable of "transforming, generating and recalling symbolic, non-linguistic information" (Linn & Petersen, 1985, p. 1482). This skillset is important for everyday activities such as orientation and navigation in one's environment, assembling furniture as well as for the performance of professional occupations like flying an airplane or carrying out scientific research (Hegarty & Waller, 2005). Miller et al. (1988) reveal that problem-solving skills improve while training spatial thinking. Equally, these abilities have a strong influence on mathematic performance (Arcavi, 2003). Furthermore, Baumert, Evans, and Geiser (1998) explain how different daily activities can improve spatial thinking abilities. There are various testing methods for evaluating spatial abilities. Early spatial ability tests were used to predict an individual's aptitude for technical occupations. These tests are based on assembling or manipulating objects, e.g. the test of tapping and aiming (Hegarty & Waller, 2005; Smith, 1964). In further studies, a large number of test set-ups are provided in order to identify different spatial ability factors, such as the cube orientation test for analyzing spatial relations and orientation. Participants were asked to decide whether two rotated cubes are identical or not in this test. In contrast to that, the paper folding test evaluates spatial visualization abilities by choosing between five alternatives of a folded and perforated paper that fits an initial figure (Ekstrom, French, Harman, & Dermen, 1976). Further studies reveal that different testing methodologies indicate varying degrees of gender differences. The test with the largest gender gap was found to be the mental rotation test (Halpern et al., 2007; Stumpf, 1993). This phenomenon could be explained by varying interests, specific brain systems as well as a wide range of cultural factors like educational influences, training and practical experience (Halpern et al., 2007). The preferable mental rotation testing procedure is the mental rotation test by

Vandenberg and Kuse, which is an advancement of Shepard and Metzler's method (Caissie, Vigneau, & Bors, 2009). The test consists of five sets with four items. In each set, there is a given sample drawing of a three-dimensional figure, which has to be recognized in the other four provided illustrations. Two of these depictions are equal to the sample but rotated. The other two pictures are modified in some details and are therefore not identical to the sample. The participant is then asked to identify the correct versions. This requires a mental rotation process and thus the ability to virtually manipulate three-dimensional images (Hegarty, 2018; Vandenberg & Kuse, 1978). According to Masters and Sanders (1993) and Vandenberg and Kuse (1978), the evaluation of several mental rotation test results shows that men score in general higher than women. Hegarty (2018) has confirmed these results in a more recent study.

Spatial thinking is an essential skill for understanding and creating instructions. In order to be able to assemble an object, it is necessary to have a graphic image of the construction in one's head. As a consequence, instructions should provide three-dimensional illustrations as well as verbal descriptions (Daniel & Tversky, 2012). In this regard, it should be pointed out that a gender difference does not exist in verbal ability, which is important for writing instructions (Hyde & Linn, 1988).

Daniel and Tversky (2012) conducted an experiment in which the participants were supposed to assemble a TV stand. This was followed by the task of creating different kinds of instructions. The design of verbal instruction with and without constraints was the first assignment. The second included designing verbal and diagrammatic instructions in the context of the mental rotation test results. The final task included the production of concise and constrained instructions with verbal and diagrammatic elements. The findings show that participants with a high spatial ability performed especially well in both diagrammatic and verbal instructions. These verbal instructions are characterized by "better and more accurate verbal instructions, including more comprehensive descriptions of the actions to be performed and fewer errors" (Daniel & Tversky, 2012, p. 317). Furthermore, the study reveals that for both, designers and users, the combination of text and diagrams is the most desired kind of instruction.

Another study from Wiking et al. (2016) views the assembly performance in context with differences between the genders. One group had to assemble a TV stand with the help of step-by-step instructions, the other group with only a picture of the final product. In addition, a MRT test was conducted. The results show that men performed faster and more accurately in the assembly process than women. Moreover, the more time was used on reading and understanding the instructions, the greater was the negative impact on the results of the MRT test.

3 Research Question and Hypotheses

As highlighted in the previous section, research has already focused on the topics of assembly, instruction design and spatial thinking. Daniel and Tversky (2012) have focused on the creation of instructions while taking spatial thinking into account. Wiking et al. (2016) have also considered gender differences regarding performance in the assembly of objects. By combining the gender differences during the assembly process and the analysis of technical understanding and spatial thinking, this study examines the question of how men and women approach and solve technical tasks. In order to answer this question, the following three hypotheses will be evaluated:

Hypothesis 1: Men have better spatial thinking abilities and therefore create better instructions than women. According to Vandenberg and Kuse (1978) and Masters and Sanders (1993), men score higher in mental rotation tests. Furthermore, studies show that people with better spatial thinking abilities create better instructions (Daniel & Tversky, 2012). Thus, it will be evaluated if there is a causal relationship between spatial thinking abilities and the creation of instructions.

Hypothesis 2: As already mentioned in the literature review, the higher the spatial abilities of a person, the more advanced are the before mentioned everyday skills (Hegarty & Waller, 2005). Simultaneously, men outperform women in spatial thinking and mental rotation tests. Thus, the hypothesis can be derived that men have a better basic technical understanding than women.

Hypothesis 3: It can be determined that there is a gender difference in risk aversion concerning financial decisions. Women are more likely to react risk-averse, while men are more ready to take risks. In this context, it is an aim to examine if women also show more risk-averse behavior in approaching and solving problems within the scope of this experiment (Borghans, Heckman, Golsteyn, & Meijers, 2009).

4 Methodology

In order to answer the research question that was outlined above and to evaluate the corresponding hypotheses, the following experiment, consisting of two parts, was designed. The first part is a quantitative-descriptive analysis and the second part includes two quasi-experiments. The whole experiment was conducted, as already mentioned in the introduction, within the context of the "International Girls' and Boys' Day" at the tech-incubator Zollhof, Nuremberg. The twelve participants of the experiment, six women and six men, are pupils of the FOS/BOS Erlangen, a secondary vocational school. The age of the participants ranged between 16 and 20 years. All participants were native German speakers. In the second part, the two quasi-experiments were recorded with a video camera in order to enable a detailed analysis of the experiments afterwards.

Quantitative-Descriptive Analysis:

The quantitative-descriptive part of the experimental setup consisted of a questionnaire and a shortened version of an MRT-Test. The seven questions of the questionnaire are derived from the study of Wiking et al. (2016) with the aim of testing past practical experiences that are associated with spatial thinking abilities.

The questionnaire was in German. The applied rating scale is the Likert scale, which is the most commonly used, one-dimensional scaling system in the field of social- and economic sciences for measuring attitudes (Albers, Klapper, Konradt, Walter, & Wolf, 2009). The applied scale entails the numbers one to six, which makes a neutral choice impossible.

The shortened MRT test consists of the first four questions of the MRT test of Vandenberg and Kruse which was already explained in the literature review. The scoring system of Vandenberg and Kruse was applied in this work as well. The questionnaire and the MRT test were handed to the pupils before executing the quasi-experiments. Furthermore, there was no explicit time limit.

Quasi-Experiments:

The second part of the experiment consisted of two empirical quasi-experiments that will be explained in the following. Each of the twelve participants took part in both experiments, one at a time. The first quasi-experiment was the assembly of a chair without instructions. The test subjects were instructed to assemble a desk chair consisting of nine different parts without any time limit. They were told to inform the staff when they completed the assembly task. The single parts of the chair were stored in a closed cupboard box. Additionally, the participants were informed that no questions would be answered during the assembly process and no further instructions on how to build the chair were provided. After finishing the task, the participants were asked to test the chair to ensure that it was stable and functionality was given. The time spent on assembly was measured with the stopwatch function of a mobile phone. The participants did not know that time was being measured.

The second quasi-experiment was the task of writing an instruction on how to assemble the chair. The participants received the following assignment:

> Imagine you have to describe the assembly of the chair. Please create an instruction that entails text and diagrams so that another person can easily and effectively build up the chair with your instruction. The diagrams don't have to be a perfect representation; it is more about good comprehensibility. You can use all the materials that you can find in the box.

By giving these instructions it was ensured that the task was well understood and that there was no pressure to perform a perfect drawing/diagram. Furthermore, the participants were informed that they can use all available materials, e.g. colored pencils, a set square, pens, household labels, a divider etc. A paper in A4 was handed out for writing the instruction. The time limit was nine minutes. After eight minutes, the participants were informed that they had one minute left to accomplish the task.

5 Results

For evaluating the results of the three experiments, there are defined suitable characteristics for each of them. Furthermore, all the characteristics were ranked on a scale from zero to four, four being the best and zero the worst rating. By using an identical scale for every measurement, comparison among all characteristics is made possible.

Five characteristics were measured for experiment one, the assembly of the chair without instructions. The logical assembly order determines whether the participant is able to develop an appropriate strategy for assembling the chair. There is only one possible way to build up the chair fast and safely. The recognition of physical connections represents the number of physical connections of the chair, e.g. plug- or click-connections, that are identified and used in the right way by the participants. Although there is no explicit time restriction, the time for assembly is measured and less time is rated higher in the evaluation. Whether a participant checks all assembled parts to ensure safety and functionality is also a measure that is applied. Experiment two, creating instructions for the chair assembly, contains seven items that are evaluated. The first one considers the completeness of the instructions. It is important that every step for assembling the chair is mentioned and described with verbal and diagrammatical elements. Therefore, the structure of the instruction is also of interest. The single steps should be identifiable and in a logical order. The allocation of text and diagrams should be clearly visible. The number of pictures included in the instruction is also measured since the understandability of the instructions is positively related to the number of pictures. The wording that is used in the instructions indicates whether a participant knows the individual components of the chair or not. Thus, this characteristic is also related to the assembly process. The perspectives that were chosen for the drawings, as well as the use of 2D/3D drawings shows the ability of the participant to change perspectives if it is required for the respective step.

Figure 6.1 shows the average score of women and men for each measured characteristic in a network diagram. The characteristics for experiment 1 are listed on the right side and the ones for experiment 2 on the left. The mental rotation test results at the top represent the results from the quantitative-descriptive analysis. It becomes obvious that the results of men and women differ in a fundamental way. Men tend to have a stronger aptitude for the characteristics from experiment 1, while women performed better in the measured characteristics of experiment 2.

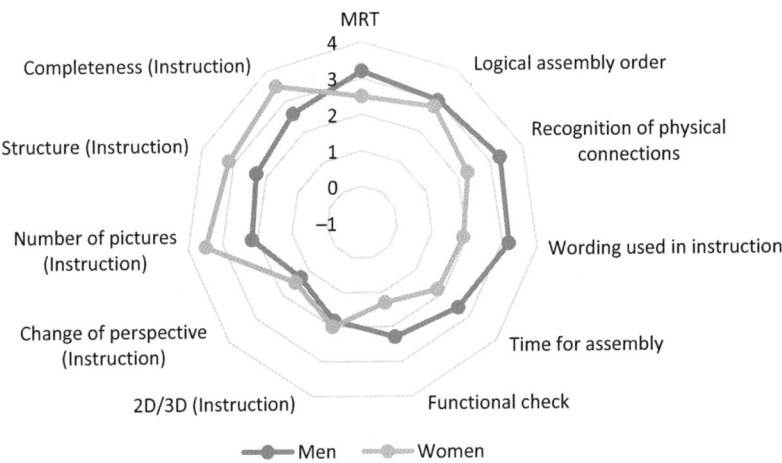

Figure 6.1: Network diagram of the measures that were tested in the experiments (Source: Own illustration).

The detailed results are presented in the following. As it can be seen, there is a strong tendency that men perform better in the assembly process than women. The biggest sex differences are identifiable in the wording that was used for the elements. The wording of men was more precise and explicit and matches the technical terms in most of the cases (Men: 3.2; Women: 1.9). The functionality check (Men: 2.3; Women: 1.3) and recognition of physical connections (Men: 3.3; Women: 2.3) shows a distinct gap. Both men and women neglected functional checks in many cases which led to an overall low score. The average time needed for the assembly of the chair amounts to less than two minutes for men whereas women needed 20 seconds more on average (Men: 2.6; Women: 1.8). Both sexes achieved almost the same results concerning the logical order of assembly (Men: 3.0; Women: 2.8). Women achieved overall better results in creating the instructions. Especially the high number of pictures women used in their instructions enhances the quality compared to men's instructions (Men: 2.2; Women: 3.5). Similar results can be seen in the completeness (Men:2.6; Women: 3.5) and structure of the instructions (Men: 2.3; Women: 3.2). Almost all instructions created by women were complete and well-structured whereas some of the men's instructions were incomplete and the structure did not guarantee an easy understanding. Neither men nor women used different perspectives in their drawings to show each of the assembly steps in an appropriate way (Men: 1.3; Women: 1.5). The results for the usage of three-dimensional drawings are almost the same for both genders (Men: 1.8; Women: 2.0). Only a few participants tried to increase the value of their instructions by using 3D-drawings. The last characteristic is represented by the results of the mental rotation test. In this test, men slightly outperform women (Men: 3.2; Women: 2.5).

Since there is no apparent difference between the genders for 2D/3D drawings and change of perspective, another analysis is supposed to evaluate a potential cause for the results. Therefore, displays all characteristics that are used for analyzing the quality of the instructions in comparison with the MRT-results. Participants with good MRT results scored at least four in the MRT-test; participants with bad results scored three or less. In contrast to , completeness, structure and the number of pictures does not show any conspicuousness. Greater dependencies can be recognized for the two visual characteristics. For a change of perspective (good MRT results: 2.0; bad MRT results: 0.8) as well as for 2D/3D drawings (good MRT results: 2.7; bad MRT results: 1.2) there is a clear gap between participants with good and bad mental rotation abilities. This big difference indicates that mental rotation abilities are the main cause for the quality and the utilization of visual characteristics.

Overall, only one male participant succeeded in assembling the chair completely and without any mistakes. Two participants were able to build up the chair with the help of hints. The majority did not finish the assembly completely, even with support.

6 Discussion

This chapter aims to identify differences between the genders in approaching and solving technical tasks. The goal is to verify the hypotheses of previous researches, like Daniel and Tversky (2012), Vandenberg and Kuse (1978) and Wiking et al. (2016) and to analyze the hypotheses which are presented in Section 6.3. The latter will be discussed in the following.

Hypothesis 1 is a combination of two hypotheses which have already been proven to be true in previous research. The first one is that men have better abilities in spatial thinking than women. That is why the mental rotation test by Vandenberg and Kuse is used since mental rotation is important for the task in the experiment. Furthermore, mental rotation abilities show the greatest differences between the genders in the field of spatial thinking. This hypothesis can be supported within our experiment since men performed up to 18% better than women. A meta-analysis of Masters & Sanders (1993) has produced similar results by identifying a variability of performance of up to 16% that was exclusively due to gender. Even though the results in Section 6.5 seemingly confirm previous research, it is important to keep the sample size and extent in mind. The second part of the hypothesis states that men create better instructions because of their advantages in spatial thinking. To verify this statement, it needs to be analyzed, whether the participants with good mental rotation test results also performed better in creating the instructions. The factors completeness, structure and number of pictures show similar results, without evidence of correlating with the MRT-results. In contrast to that, Figure 6.2 shows a

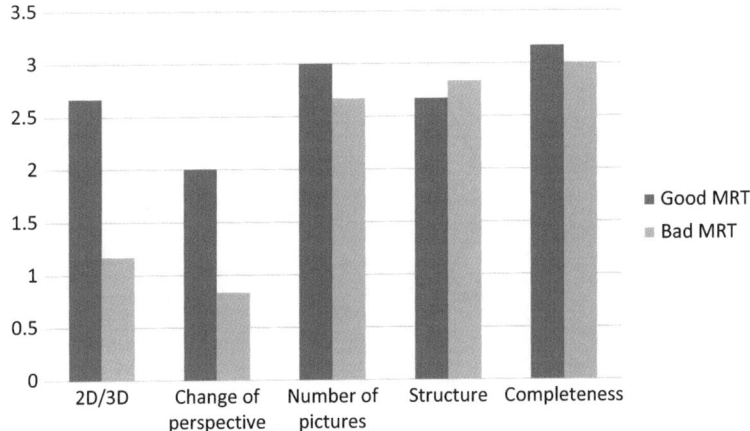

Figure 6.2: Characteristics of the instructions in connection with the MRT-results (Source: Own illustration).

clear dependency between the measured characteristics and the MRT results. Participants with good MRT-results scored 37% better in 2D/3D drawings and 29% better in changing their perspective than participants with bad results. Thus, the instructions of participants with good spatial thinking have the potential to be better than the instructions from their peers with less developed spatial abilities. Daniel & Tversky (2012) revealed a correlation of spatial ability and the creation of instructions not only for the diagrammatical part but also for the verbal part of the instructions. For validation of the final hypothesis, it is necessary to check the influence of gender on the quality of instructions. Women's instructions were overall more complete, better structured and contained more pictures. Only for 2D/3D-drawings and the change of perspective, there is no remarkable difference. In summary, the hypothesis can partly be confirmed since indeed, men scored higher in the MRT-test and spatial thinking abilities have a significant influence on the quality of instructions. Nevertheless, women create better instructions and therefore, the initial hypothesis that men create better instructions because of their advanced spatial thinking skills can partly be disproved.

As already explained previously, researchers have revealed a significant influence of spatial thinking abilities on everyday activities, such as orientation or catching a ball (Hegarty & Waller, 2005). This research chapter aims to investigate the relationship between gender, spatial thinking and technological understanding. Therefore, the hypothesis that men have a better basic technical understanding than women must be tested. The characteristics logical assembly order, recognition of physical connections and wording are relevant for this area. While there is no recognizable difference between the genders in the assembly order, a gender gap becomes visible for the recognition of connections and wording. The fact that men handle the physical connections more precisely indicates that they are used to dealing with technical items and therefore know how to use them. The more appropriate wording that men used in their instructions shows higher levels of interest and experience. Thus, this hypothesis can be confirmed within our research. Even though spatial thinking might be a reason for the differences, a bigger context should be applied for an explanation. Peters et al. (1995) and Wiking et al. (2016) tried to establish a connection between spatial thinking and the activities that the participants performed in their childhood. However, a qualitative analysis of the participants' previous activities and comparison with the results from the experiment did not reveal any dependencies. This may be because of the sample size.

Hypothesis 3 evaluates if women are more careful and risk-averse than men. The assembly time and the degree to which the functionality was checked are relevant measures here. Borghans et al. (2009), as well as Jianakopolos and Beransek (1998) revealed that men are in general more prone to taking risks and thus need less time for making decisions in ambiguous situations. In this experiment, women needed more time for assembling the chair than men, which supports the hypothesis. Furthermore, it is assumed that women tend to invest more effort into checking their own decisions. This is evaluated by reviewing the assembly behavior. Within our experiment, women checked the functions of the chair very rarely, whereas men executed more functional checks. Hence, this hypothesis is partly disproven within our research because women needed more time to make decisions on the one hand but on the other hand did not adequately check their actions.

This research provides two main conclusions: Men assemble the chair faster, more precisely and efficiently. Women demonstrate their overall understanding of the assembly process by creating instructions of higher quality that are more complete and structured. Besides the mentioned explanations, such as spatial thinking and previous activities, the physical abilities of women must also be taken into account. Especially for connecting some of the elements of the chair, it is necessary to use a minimum of power. Some of the female participants could not supply this power and therefore failed the assembly process. It can be concluded that only by combining the abilities of men and women, overall excellent results can be achieved. In a broader context, the collaboration of women and men in STEM-related jobs is essential for ideal problem-solving.

7 Research limitations and implications

In this section, a critical review of the results and an outlook for future research are presented. The biggest limitation in this study was the sample size, the number of participants that were tested. The number of twelve participants is very low and cannot provide a wholesome view of gender differences in technical problem-solving. In this context, it has to be taken into account that the experimental projects of six student groups participating in the seminar had to be completed within the one day of the "International Girls' and Boys' Day". Nevertheless, future studies could be conducted in the same way but with a much bigger sample size. Moreover, the target group of the study could be changed to students who are doing STEM-related studies and another group that is doing social or economic studies. This would lead to a wider picture regarding the differences in the groups of gender itself and a better comparison would be possible. Another interesting approach is to give the designed instructions to another group of participants that have no prior knowledge of the experiment. They could be given the task to decide which of the instructions explain the assembly process best. Thereby it could be evaluated if men, for example, would prefer the instructions made by women. This would support the finding within the presented study that women have better skills in writing instructions for technical tasks.

In addition, the design of the study could be conducted with a steadily increasing number of participants. Quasi-experiments were applied within the scope of this study. These are empirical studies that don't include randomization, which means that internal validity is very limited. A change of the depending variable, for example, can't fully be traced back to the independent variable and therefore the study group may provide weaker evidence. A control group for women and men was also not implemented in this study. This leads to the limitation of not having the possibility to compare the results to a gender-specific baseline. The study could, therefore, be extended with two control groups which have to build up the chair using instructions. This would have the advantage of stronger casual relationships and the determination of more precise conclusions.

Furthermore, the focus of this study was on the process of creating instructions for assembling a desk chair. The number of parts that had to be used for assembly was comparatively low in order to measure gender differences in technical problem-solving. Thus, a different assembly task with a higher number of assembly steps could be chosen. But also the physical strength that was needed for the assembly process has to be taken into account. As already mentioned in the discussion, women had big problems in the assembly because of the weight of some of the parts and because of stiff connections. Future research could use an assembly task that is less strongly connected to physical strength. An interesting example for such a task could be an assembly process that uses computer simulation or a video game. This could eliminate the above mentioned possible disadvantages for women regarding physical strength or over-carefulness during the assembly process.

Other skills that might be connected to technical problem solving could be identified in further studies in order to get a broader picture of possible sex differences in technical problem-solving.

References

Albers, S., Klapper, D., Konradt, U., Walter, A., & Wolf, J. (2009). Methodik der empirischen Forschung. Wiesbaden: Gabler Verlag.

Anger, C., Berger, S., Koppel, O., & Plünnecke, A. (2017). MINT-Herbstreport 2017: MINT und Digitalisierung - Herausforderungen in Deutschland meistern. *Institut Der Deutschen Wirtschaft Köln*.

Arcavi, A. (2003). The role of visual representations in the learning of mathematics. *Educational Studies in Mathematics, 52*(3), 215–241.

Baumert, J., Evans, R. H., & Geiser, H. (1998). Technical problem solving among 10-year-old students as related to science achievement, out-of-school experience, domain-specific control beliefs, and attribution patterns. *Journal of Research in Science Teaching, 35*(9), 987–1013.

Borghans, L., Heckman, J. J., Golsteyn, B. H. H., & Meijers, H. (2009). Gender Differences in Risk Aversion and Ambiguity Aversion. *Journal of the European Economic Association, 7*(2–3), 649–658.

Caissie, A. F., Vigneau, F., & Bors, D. A. (2009). What does the Mental Rotation Test Measure? An Analysis of Item Difficulty and Item Characteristics. *The Open Psychology Journal.* (2), 94–102.

Daniel, M.-P., & Tversky, B. (2012). How to put things together. *Cognitive Processing, 13*(4), 303–319.

Dörner, D. (1984). Denken, Problemlösen und Intelligenz. *Psychologische Rundschau, 35*(1), 10–20.

Duncan, R. B. (1972). Characteristics of Organizational Environments and Perceived Environmental Uncertainty. *Administrative Science Quarterly, 17*(3).

Ekstrom, R. B., French, J. W., Harman, H. H., & Dermen, D. (1976). Manual for the Kit of Factor-Referenced Cognitive Tests. *Educational Testing Service*.

Halpern, D. F., Benbow, C. P., Geary, D. C., Gur, R. C., Hyde, J. S., & Gernsbacher, M. A. (2007). The Science of Sex Differences in Science and Mathematics. *Psychological Science in the Public Interest : a Journal of the American Psychological Society, 8*(1), 1–51.

Hegarty, M. (2018). Ability and sex differences in spatial thinking: What does the mental rotation test really measure? *Psychonomic Bulletin & Review, 25*(3), 1212–1219.

Hegarty, M., & Waller, D. A. (2005). Individual Differences in Spatial Abilities. In P. Shah & A. Miyake (Eds.), *The Cambridge Handbook of Visuospatial Thinking* (pp. 121–169). Cambridge: Cambridge University Press.

Hyde, J. S., & Linn, M. C. (1988). Gender differences in verbal ability: A meta-analysis. *Psychological Bulletin, 104*(1), 53–69.

Jianakopolos, A., & Beransek, A. (1998). Are Women More Risk Averse? *Economic Inquiry, 36*(4), 620–630.

Linn, M. C., & Petersen, A. C. (1985). Emergence and Characterization of Sex Differences in Spatial Ability: A Meta-Analysis. *Child Development, Vol. 56*, 1479–1498.

Marples, D. L. (1961). THE DECISIONS OF ENGINEERING DESIGN. *IRE Transactions on Engineering Management, EM-8*(2), 55–71.

Masters, M. S., & Sanders, B. (1993). Is the gender difference in mental rotation disappearing? *Behavior Genetics, 23*(4), 337–341.

Miller, R. B., Kelly, G. N., & Kelly, J. T. (1988). Effects of Logo computer programming experience on problem solving and spatial relations ability. *Contemporary Educational Psychology, 13*(4), 348–357.

Peters, M., Laeng, B., Latham, K., Jackson, M., Zaiyouna, R., & Richardson, C. (1995). A redrawn Vandenberg and Kuse mental rotations test: Different versions and factors that affect performance. *Brain and Cognition, 28*(1), 39–58.

Savransky, S. (2000). Engineering of Creativity: Introduction to Triz Methodology of Inventive Problem Solving. CRC Press.

Schrader, S., Riggs, W. M., & Smith, R. P. (1993). Choice over uncertainty and ambiguity in technical problem solving. *Journal of Engineering and Technology Management, 10*(1–2).

Smith, I. M. (1964). Spatial ability: Its educational and social significance. London: University of London Press.

Stumpf, H. (1993). Performance factors and gender-related differences in spatial ability: Another assessment. *Memory & Cognition*, 828–837.

Vandenberg, S. G., & Kuse, A. R. (1978). Mental rotations, a group test of three-dimensional spatial visualization. *Perceptual and Motor Skills, 47*(2), 599–604.

Wiking, S., Brattfjell, M. L., Iversen, E. E., Malinowska, K., Mikkelsen, R. L., Røed, L. P., & Westgren, J. E. (2016). Sex Differences in Furniture Assembly Performance: An Experimental Study. *Applied Cognitive Psychology, 30*(2), 226–233.

Stephanie Birkner, Janina Sundermeier, Silke Tegtmeier
7 E-health Value Creation Revisited: Towards a Gender-Aware Typology of Digital Business Models

1 Introduction

Health is a fundamental human right and defined by the World Health Organization as the "state of complete physical, mental, social and spiritual well-being and not merely the absence of disease or infirmity" (WHO 1946, cited in Dhar et al., 2011). Since the 18th century, sanitary health legislation, disease control, and epidemiology have brought substantial improvements to public health and the quality of life (for its historical roots see e.g. Porter, 1999).

However, health is not equally available across the world population and some countries and social groups suffer significant health deficits. These deficits cause enormous economic and social costs, damages and losses (OECD, 2017). Indeed, while the history of medicine is a true success story, society has now reached a state where the current progress is not enough to tackle the postmodern demands of physical and psychological well-being (e.g. demographic change, multi-drug resistance [MDR], or increasingly common mental health issues). One approach to counteract these developments is to address the issues around gender (in-)equality, in order to increase value creation options in (public) health, since the negligence of these topics in the past has not only been a costly and unacceptable social issue but also a missed market opportunity (EC, 2013).

Márie Geoghegean-Quinn, European Commissioner for Research, Innovation and Science from 2010 to 2014, sums up the scientific discourse on the innovation potential of gender analysis as follows: "Gender Analysis contributes to excellence; it stimulates new knowledge and technologies; opens new niches and opportunities for research streams and results in products and services that all members of our society need and demand" (EC, 2013, p. 6).

Gender analysis is particularly relevant to research processes and aims to turn scientific evidence into new business models. A significant amount of research has already been undertaken concerning the integration of gender into health research processes in general, from formulating research questions, to designing methodologies and interpreting related data (Schiebinger, 2014b). However, little is known to date about how these practices and related findings are transferred to the business sector through academic outreach activities (e.g. knowledge transfer or research impact).

Alsos et al. (2013) discuss several combinations of perspectives on gender (as variable, construction, and process) and innovation (as result, process, and discourse/

policy) that give rise to multiple and heterogeneous implications for practice and policy.

In the current chapter, we explore the issue of how gendered innovation potential can be exploited for the benefit of value creation. Our topic lies at the intersection of "gender as a construction" and "innovation as result" and discusses insights from entrepreneurship research. Specifically, we address this research gap by conceptualizing a heuristic classification scheme to explore the extent to which business models address (or ignore) gender. Our focus is especially on digital business models. Indeed, given the rise of information technology in the 20th century and the socio-economic digital transformation of the 21st century, the landscape of public health, and especially health care, is being fundamentally transformed. To that end, health services and information are delivered and/or enhanced through digital technology that have given rise to "e-health" solutions. E-health is an emerging research field at the intersections of public health, medical informatics and entrepreneurship (Eysenbach and Jadad, 2001). Developments in the research field of e-health are promising (e.g. Eland-de Kok et al., 2011) and the number of applications is growing rapidly (Mair et al., 2012). Numerous innovations, such as smart health monitoring systems (SMHS) or digital patient-reported outcomes (PRO), have facilitated value creation in e-health in terms of improved quality and efficiency of care, including clinical trials and clinical decision-making (e.g. Valderas et al., 2008; Haverman et al., 2011; Wicks et al., 2011; Haverman et al., 2012; Paton et al., 2012; Snyder et al., 2012).

Interestingly, e-health and gendered innovations are rarely linked, disregarding the potential of synergetic co-creation of gender-aware (public) e-health value propositions. Insights from entrepreneurship research are also rarely used to innovate at the interface of (societal) needs, (health) market impact and technological progress – except for the (physical and mental) health of entrepreneurs (Shepherd and Patzelt, 2015; Stephan, 2018). Research on business models as a source of value creation is only at the beginning of being included in healthcare research (e.g. Brady and Saranga, 2013; Fredriksson et al., 2017; Winterhalter et al., 2017).

Scholars have started to survey the concept of business models from various theoretical perspectives as a distinct unit of analysis (Osterwalder et al., 2005; Casadesus-Masanell and Ricart, 2010; Teece, 2010; Amit and Zott, 2012; Spieth et al., 2014), which might provide reasonable insights for studying value creation in (e-)health. In light of these considerations, the contribution of this chapter is threefold: (1) We extend the existing evidence concerning the relevance of gendered innovation for research endeavours to the area of entrepreneurship research; (2) We want to extend the knowledge on business models to research that lies at the intersection of digital transformation and innovation needs and potentials in (public) health; (3) We conceptualize a heuristic scheme that paves the way for further research into how gender-awareness is manifested in digital business models.

2 Creating Value at the Crossroads of Digital Transformation and (Public) Health

2.1 Relevance of Gendered Innovation for Endeavors in Entrepreneurship Research

While the European Commission (EC) has emerged as the global leader of gender-aware research policies (e.g. EC, 2013, 2003a, 2003b), gender-awareness has also found its way into the research agendas of, for example the Canadian Institutes of Health Research (CIHR, 2010) or the US National Institute of Health (Clayton and Collins, 2014). Londa Schiebinger, Professor for History of Science, Stanford University, has been at the forefront of gender-aware research. Her work has exposed the harm done by unconscious gender bias and highlighted the importance and potential of gender analysis in research which could lead to new discoveries in many sciences, including health/medicine, and engineering (Schiebinger and Schraudner, 2011; Schiebinger, 2014b; Schiebinger and Klinge, 2015). Recently, an interdisciplinary group of sixty experts directed by Schiebinger investigated the role of gender differences (in terms of needs, behaviours and attitudes) in research design/content, and thus highlighted the societal relevance and quality of research outcomes, coining the term "gendered innovations" (Schiebinger et al., 2011–2017). As such, the group has implemented the recommendation of the United Nations' Resolutions on Gender, Science and Technology as well as the genSET Consensus Report (GenSET 2010) regarding the global reach of science and technology, methods of sex and gender analysis developed through international collaborations (e.g. UNESCO's International Report on Science, Technology and Gender, UNESCO 2007).

The gendered innovations project led by Schiebinger provides scholarly, as well as practical guidance and tools. Most importantly, it contributes to the discussion regarding the distinction between sex and gender to foster the accurate use of the terms to inform research. Fishman et al. (1999), for example, highlight the difference between these two terms from the perspective of health care, specifically pain perception: Sex analysis addresses biological determinants while gender analysis addresses socio-cultural assumptions about how people experience pain differently. While analytically both terms are distinct, in the "real" world they interact in complex and thus important ways. Krieger (2003) refers to this relatedness by accentuating the "gendered expressions of biology" and "the biologic expression of gender" to discuss phenomena that are both biologically and socially constructed (Fausto-Sterling 2012). Gender aware science and technology have the potential to improve the lives of human beings worldwide. Gendered innovations refer to "processes that integrate sex and gender analysis into all phases of basic and applied research to assure excellence and quality in outcomes" (EC, 2013, p. 9). It covers all stages of a research cycle: From setting priorities for research and research funding, to formulating project objectives, developing methodologies, gathering and analysing data, evaluating results, developing patents, transferring ideas to markets, and drafting policies.

Gendered innovations are intended to enhance creativity, innovation, and gender equality (Schiebinger, 2014b). Corresponding guidelines and tools are designed to add value to (1) research and engineering (to ensure excellence and quality in outcomes, and enhance sustainability), (2) the society in general, by making science more responsive to social needs, and (3) businesses, by fostering innovation through patents and new technologies (Schiebinger and Schraudner 2011). This is especially relevant for the (public) health sector, because the absence of gender-awareness in research can have serious and even life-threatening consequences (Schiebinger, 2014b). Similarly, this can cause delays in diagnosis and treatment in a care context (Chilet-Rosell 2014; EC 2013).

Despite the growing recognition of gender analysis that focuses on sex and/or gender as a central factor to generate the best possible research evidence for the health sector (Doyal, 2001; Bierman et al., 2007; Greaves, 2011; Coen and Banister, 2012), the integration of gender analysis into health research has been slow and uneven (Day et al., 2016), both in regards to the design and the reporting of health studies (e.g. Gahagan et al., 2015; Lee et al., 2018). Researchers have only recently started to gather and analyse data worldwide to gain insights into the relationships between healthy behaviours and outcomes and gendered social norms/gender inequalities (Darmstadt et al. forthcoming). However, the question of how these insights can be transferred to socio-economic value creation has so far remained unanswered. With our approach we propose to link aspects of digital transformation to foster gender-awareness in health entrepreneurship.

Unlike any previous basic technological innovation (see Joseph A. Schumpeter, 1961, for the "bandwagon effect" of basic innovation), information and communication technologies (ICT) have fostered socio-economic transformations during the second half of the 20th century. From an economic perspective, "countries in the advanced digitization stage reap 20 percent more economic benefits than countries at the start of their digitization journeys" (Gulati and Soni, 2015, p. 60).

According to Nefiodow and Nefiodow (2014), the entropic sector plays a key role in the economy, because the enormous economic losses, damages and costs incurred year after year in this instance have turned this into a significant barrier to economic and societal development. Wick et al. (2014) state that digital health care innovations provide great opportunities to decrease costs and at the same time increase productivity (Wicks et al., 2014).

2.2 Value Creation at the Crossroad of Digital Transformation and (Public) Health

In both research and practice, digitalization and its power to transform and disrupt has taken centre stage in academic publications, scientific panels and people's concerns. Digitalization as a megatrend has been attracting scholarly attention in several disciplines as it is considered a source of and a driver for our turbulent and disruptive era, characterized by cultural shifts, blurred industry boundaries, and new forms of (collaborative) working and living.

In fact, the story of digitalization is not as new as one may think. Its roots can be traced back to the science of cybernetics ('control theory') in the mid-20th century, focusing on the analysis of control and communication aspects between humans and machines (e.g. Wiener, 1948; Ashby, 1966). With the advent of the internet (aka the "worldwideweb") in the 1990s, firms have been challenged to adhere to the evolving demands of digitizing the creation, distribution and management of their value proposition; Digitization had become a key part of the business strategy (Gulati and Soni, 2015). The 1990s became famous for the economic growth resulting from the use and widespread adoption of the internet followed by the historic economic crash of the early 2000s (Howcroft, 2001; Valliere and Peterson, 2004). Still, economic life after the dot.com bubble continues to revolutionize business practices (Min et al., 2008). Indeed, the internet – and digital technologies of all kinds – have become an integral part of daily life and work routines. The extent of digitization is such that its perception and utilization has come to define the generation one belongs to and created generational divides between so-called "digital natives" and "digital immigrants" (Prensky, 2001; see Judd, 2018 for a critique).

The ongoing creation of digital versions of objects ('digitization') has led to leveraging their application which does not stop with the digital objects themselves but affects the systems and processes to which they are connected. This "digitalization" – the altering of (socio-) economic processes during the process of adapting to the challenges and chances of a digitalized world – has created a "digital shift". The (re-)design of workplaces or supply chains, or even of complete business models, is one of its manifestations. However, the adaptation of business structures and processes in turn invites and requires a reassessment of approaches towards living and working in the digital era. Nevertheless, the term "shift" implies that a new level is reached with profound effects but also unknown consequences. Given the increasing insights on and the potential of digital technologies with respect to not only human-machine interactions (Marquez et al., 2018) or human-machine cooperation (Hoc, 2000) but also interactions of all kinds of living beings with digital technology (e.g. Wirmanm, 2014), another term is required to describe forms of interactions between non-digitized and digitized entities at all levels. For example, digital technologies like Artificial Intelligence (AI) are considered to be transforming our daily lives (e.g. Fox, 2018). In summary, it can be said that the "digital era" is characterized by a journey from digitization towards digital transformation – or from the creation of digital content to transformations caused by (and in support of) digitalized structures and (inter-action) processes. These changes occur in the intertwined realms of the economy, natural environment, and society. Table 7.1 summarizes and contrasts the differences between the definitions of digitization, digitalization, digital shift, and digital transformation.

Table 7.1: Contrasting digitization, digitalization, digital shift and digital Transformations (Source: Own Illustration).

Verb	Noun	Modus Operandi	Definition
to digitize	digitization	singular	creating digital version of objects
to digitalize	digitalization	linked	leveraging digitized objects through system connections
to digitally shift	digital shift	joined	altering (socio-)economic processes due to digitalization
to transform because and in support of digitalization	digital transformation	holistic	disrupting continuously appearance and procedure of forms of interactions between non-digitized entities (N) and digitized entities (D)

According to the WHO (2009), the scope of socio-economic innovations that address aspects of (public) health is holistic. In this context, the intersection of digital technologies and innovation/entrepreneurship provides ample scholarly as well as practical opportunities for value creation in the health sector, given that digitization transforms innovation processes and outcomes, as well as modes of "traditional entrepreneurship" (Nambisan et al., 2017; Nambisan, 2017). In the function of external enablers, digital technologies are able to stimulate and foster new venture creation processes (von Briel et al., 2018) in the form of, e.g. digital applications, components, media content (Ekbia, 2009), products (Lyytinen et al., 2016), and platforms (Tiwana et al., 2010).

The emerging intersection of (public) health, entrepreneurship and (medical) computer sciences is referred to as electronic health (e-health). Benefits of this interdisciplinary intersection of research and practice are assumed to be manifold, e.g., through (1) increasing efficiency/lowering costs by avoiding duplications of diagnostics and therapeutic interventions, (2) improving quality by, e.g. including patient reports, and (3) empowering individuals on the basis of clinical, as well as peer-to-peer induced health literacy (Wicks et al., 2014). Table 7.2 provides examples for all four digital modus operandi (digitization, digitalization, digital shift, and digital transformation) in connection with healthcare practices supported by electronic processes and communication (e-health).

Table 7.2: E-Health examples of digital modus operandi (Source: Own Illustration).

Modus Operandi	Example of healthcare practice supported by electronic processes & communication (e-health)
singular (digitization)	diagnostic sonography (ultrasound contrast imaging)
linked (digitalization)	psychophysiological assessment (digital technology supported data collection and analysis of autonomic nervous system and the neuroendocrine system)
joined (digital shift)	telemedicine (remote diagnosis and treatment of patients)
holistic (digital transformation)	augmented surgery (CAMI – computer-assisted medical intervention)

Given the relevance of both gender analysis and digitalization for health research and practice, it is surprising that knowledge of how gendered innovations are reflected in recent (digital) business models is still scarce due to a lack of empirical evidence in this area. From a scholarly perspective, such knowledge is essential to determine the impact of gender-conscious funding frameworks in leading policy programs aimed at promoting a "digital strategy" within the health sector.

From a research perspective, three creative ideas are central to doing research: the idea of cause, the idea of chance, and the idea of order (Bronowski and Long, 1951). Especially the latter has, over the last decade, gained popularity in the science communities concerned with (innovation) management, entrepreneurship, and technology, with the conceptualization of "business models" as a tool for describing and analysing how value is created through business processes. In the wake of the start-up hype in social media, and TV programs such as "Startup" or "Dragon's Den", the term "business model" has even reached the general public and is in widespread use. Despite this popularity, it is notable that there remains a certain reluctance in academia to put the concept itself at center stage and to acknowledge and discuss its history, current status and prospects. The aim of this chapter, therefore, is not to discuss business models from a general perspective, but to propose a "heuristic scheme" for the analysis of business models and thereby pave the way for developing further theoretical approaches for the analysis of business models from a gendered perspective.

3 Gender-Aware Heuristic Scheme for Researching Business Models in E-Health

The number of publications on the characteristics of business models is continuously growing, focusing, e.g. on their attributes (Dubosson-Torbay et al., 2002; Hedman and Kalling, 2003; Osterwalder, 2004; Pateli and Giaglis, 2004; Osterwalder et al., 2005) or their embeddedness in tech-ecosystems (Afuah and Tucci, 2003; Pateli and Giaglis, 2004), among others. However, business models still represent "a slippery construct to study" (Casadesus-Masanell and Zhu, 2013,p. 480) not least due to the blurred boundaries of the construct (George and Bock, 2011; Ricciardi et al., 2016; Ritter and Lettl, 2017). The lack of a common theoretical framework (Zott et al., 2011; Spieth et al., 2014) converts the scholarly call for a general classification scheme of business models more and more into a distinct plea (Hawkins, 2002; Pateli and Giaglis, 2004; Qureshi and Keen, 2005; Baden-Fuller and Morgan, 2010; Lambert, 2015).

For studying organizational systematics, McKelvey (1982) proposes in accordance with Jeffrey (1982) two basic classification approaches: The first is a general classification (which considers all attributes) and the second is a more specialized classification (only few attributes are considered). In other fields of research, the latter is also referred to as artificial/arbitrary classification (Simpson, 1961; SoKAL and Sneath, 1963). All forms of ordering cases (objects) into groups to provide meaning to reality (Simpson, 1964) based on their similarities, can be considered a form of classification (Bailey, 1994). Thus, the overall target or purpose of classification is that we do not need to consider, remember, or discuss each case as a unique entity but are able to link insights of one case to existing knowledge of other cases of its kind (Smith and Medin, 1981).

With respect to specified criteria to classify business models, numerous approaches can be found in literature studying business models as a dependent or independent variable (Zott et al., 2011). For example, Bienstock et al. (2002) examine product-related factors and market- related criteria such as customer profiles (see as well Seong Leem et al., 2004); Timmers (1998) and Tapscott (2000) research market configuration factors while Weill and Vitale (2002) and Betz (2002) focus on resources. Furthermore, special classifications of business models are conceptualized for the purpose of empirical studies comparing modes of management (Rajala and Westerlund, 2007) or the survival of firms (Kauffman and Wang, 2008). In a wider context, there is a growing discourse on ontological classifications, examining the factors that constitute business models (Hedman and Kalling, 2003; Pateli and Giaglis, 2004; Gordijn et al., 2005; Osterwalder et al., 2005).

In a comparison of business models from a historical perspective, models have been clustered according to how entrepreneurial activity was organized, or by the extent to which they have brought about ground-breaking inventions/basic innovations that have launched economic (r)evolutions (Kondratieff and Stolper, 1935; Schumpeter, 1939). In a similar vein, it could be explored if historical versions of business models (e.g. the guilds of the middle ages, the factories of the 19th century, and the networks of the 20th century) are unique to the eras in which they have arisen or whether they can be revived, and if so, under which conditions. One example is the recent renaissance of the cooperative model that is regaining popularity, especially in the energy sector (e.g. Klagge and Meister, 2018; Yamamoto, 2018).

In a general classification of historical forms of business models, all attributes of the analysed entities are potentially of interest for a better understanding of each model.

Broadly speaking, two philosophical discourses on classification can be distinguished that determine the ontological, epistemological, and methodological background of classification design. On the one hand, classifications are discussed from an essentialist perspective, as typologies. From an Aristotelian perspective, while classes can be defined based on a few essential characteristics, typologies in the sense of monothetic grouping can be derived from common sense or theory (Warriner, 1984). In contrast, taxonomies stem from a classification philosophy of empiricism, which is based on Adansonian principles grounded in the work of the 18th century naturalist Michel Adanson. The aim here is to order cases according to their degree of affinity as polythetic groups such that cases share the greatest number of characteristics but without any single characteristic being required for the case to belong to the group (Sneath and Sokal 1973).

Based on extant literature on classifications, Lambert (2015, p. 55) proposes a scheme that on the one hand "encourage[s] the application of theoretical rigor to the design of classification schemes in business model research and communicating their underlying structure to potential users" and on the other hand, "lead[s] to a classification outcome appropriate for the intended purpose". In the following, this chapter makes use of Lampert's approach to conceptualize a "heuristic frame" for entrepreneurship research to aid the survey of gender-awareness in digital business models. Thus, we aim to bring about valuable insights of how business models can be classified in general, and especially at the crossroads of digital transformation, gendered innovations, and (public) health. In particular, we propose that the purpose of the conceptualization of a classification approach for gender-aware e-health business models is distinct, and should be bounded by the following criteria: (1) The core of the business model (value proposition) has to be digital; (2) The customers of the business model have to be in the health sector; (3) The business model has to either focus on or at least integrate aspects of gender analysis.

All in all, for the development of a typology, an essentialist classification philosophy that builds upon a deductive approach appears to be most suitable. Bailey (1994) points out that typologies are in most cases developed through a qualitative rather than a quantitative analysis. However, typologies can also be compiled by conceptualizing the criteria of the classification scheme and then using empirical analyses to validate them, so that "[a] sound typology forms a solid foundation for both theorizing and empirical research" (Bailey 1994, p. 33).

The essentialist classification approach proposes that classification criteria should be derived from state-of-the art research within the field of interest (Lambert, 2015). For the proposed "heuristic scheme", this chapter focuses on literature in each separate field of interest as well as on literature that lies at the junction between business models, digitalization and gendered innovations in the field of (public) health. The insights from the literature review inform the categories with which monothetic groups (essentialist approach) are framed, as opposed to polythetic groups (empiricist approach; or as much as possible surveyed through observation).

To analyse to what extent business models manifest gender-awareness, a procedure must be developed that is consistent with the principles of the chosen classification philosophy. As discussed above, this means that the essentialist approach requires a procedure that can be used to identify a small number of criteria to conceptualize categories, while the empiricist approach needs a procedure that is conductive to the discovery variables. In respect of a heuristic scheme for a gender-aware business model classification we propose – on the basis of the literature reviewed in this chapter – to focus on criteria and their related categories: (1) gender-awareness: gender, sex, or the intersection of both terms, (2) central digital modes: digitized, digitalized, digitally shifting, digitally transforming, and (3) core business model attributes: offer (key partners, key activities, key resources), demand (customers, distribution, channels), cost and revenue streams, value proposition. Figure 7.1 visualizes the proposed

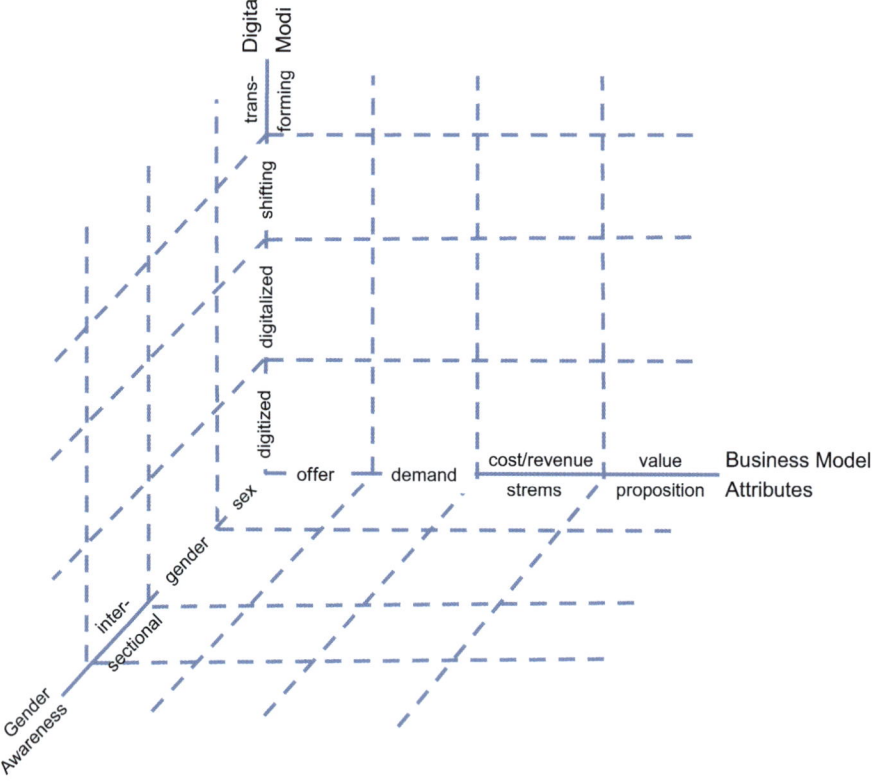

Figure 7.1: Heuristic scheme to support gender-aware business model classification (Source: Own Illustration).

deductive "heuristic scheme" which is considered to be open towards the integration of new categories that may emerge as inductive insights from future empirical surveys.

4 Discussion and Conclusion

As its central contribution, this chapter aims to conceptualize a "heuristic scheme" to analyse the gender-awareness of business models in order to advance the development of theoretical concepts. We discuss the advantages of classification schemes for the abstraction and organization of complex concepts. Overall, the classification of cases within a distinct research domain – whether to deepen knowledge about cases or to broaden knowledge about analogies to other classifications – is an important and necessary scholarly approach to understand a research area and extend its theoretical foundation. We therefore propose that the development of a classification model to identify and (re-)consider similarities and differences between business models is fundamental to business model research and forms not only an integral part of entrepreneurship and business research but also contributes to the study of business models.

To avoid misunderstandings, all researchers working on classification schemes carry out taxonomic activities in that they study classifications in respect of certain principles, rules, and procedures (Simpson, 1961). Still, the outcome of their research can either be a typology based on only a few characteristics, or a taxonomy which is the outcome of object grouping with respect to all observable characteristics.

Research on business models has, until now, predominately been conducted through neoliberal economic theoretical lenses. Given that these lenses are being criticized for ignoring gender inequality, the question begs whether they are appropriate for integrating aspects of gender analysis in entrepreneurship research in general and in digital business models, particularly in a health context. Following the plea voiced by Calás et al. (2009) for feminist discourses in entrepreneurship, we argue that future research should base heuristic schemes with a feminist focus on productive/reproductive economic issues in the care realm (see for further points of reference to feminism in entrepreneurship, e.g. Ahl and Marlow, 2012; Jennings and Brush, 2013; Muntean and Ozkazanc-Pan, 2015; Henry et al., 2016).

Methodologically, we propose to draw on a comparative case study analysis (e.g. Campbell, 2010; Bartlett and Vavrus, 2017). However, the less tangible a case (or object) is, the greater the challenge of deciding rules for sampling. Business models as research objects are quite abstract, given that there is no universally accepted definition (DaSilva and Trkman, 2014; Fjeldstad and Snow, 2018) and their perception is user-specific (George and Bock, 2011; Ricciardi et al., 2016; Ritter and Lettl, 2017). For example, value drivers can be quite complex and there is empirical evidence that a plurality of business models can co-exist in one and the same firm

(Benson-Rea et al., 2013). Given the broad range of possible e-health business model interfaces with respect to value creation in care literacy on one hand (including prevention, diagnostic, and treatment), assessment and surveillance and, on the other, the paucity of gendered innovations in practice, both single and multiple business models should be studied. Above all, the surveyed business models should express a distinct gender-awareness – not just value creation whereby lip service is paid to gender, but one where a robust gender dimension is embedded in the business model. Therefore, the classification survey should focus on startups in the post-seed phase that have already tested their prototype/minimum viable product (MVP) with a broad focus on the health market including business models addressing aspects of care literacy (including prevention, diagnostic, and treatment), and assessment/surveillance.

This chapter carries limitations in as much as typologies are good for descriptive purposes but have limited explanatory or predictive power (Hambrick, 1984). However, like theories, classification schemes that remain forever unfinished – just serving interim acceptance – can be elaborated, refined or disconfirmed by further studies (McKelvey, 1982). Of interest here would be research that empirically focuses on the coding of crowdfunding and/or crowd-financing platforms that focus on e-health value propositions (e.g. Swan, 2012).

This chapter also provides implications for practitioners which are twofold: (1) Entrepreneurs might want to make use of the classification scheme to get a deeper understanding of the content and structural choices they have to make about gender-aware value creation in e-health and as such come up with more sophisticated business models; (2) Investors and policy makers might consider the heuristic scheme as a useful lens to inform their decisions.

References

Afuah, A., & Tucci, C. L. (2003). A model of the Internet as creative destroyer. *IEEE Transactions on Engineering Management*, *50*(4), 395–402.

Ahl, H., & Marlow, S. (2012). Exploring the dynamics of gender, feminism and entrepreneurship: advancing debate to escape a dead end? *Organization*, *19*(5), 543–562.

Alsos, G. A., Ljunggren, E., & Hytti, U. (2013). Gender and innovation: state of the art and a research agenda. *International Journal of Gender and Entrepreneurship*, *5*(3), 236–256.

Amit, R., & Zott, C. (2012). Creating value through business model innovation. *MIT Sloan Management Review*, *53*(3), 41.

Ashby, W. R. (1966). The cybernetic viewpoint. *IEEE Transactions on Systems Science and Cybernetics*, *2*(1), 7–8.

Baden-Fuller, C., & Morgan, M. S. (2010). Business models as models. *Long Range Planning*, *43*(2–3), 156–171.

Bailey, K. D. (1994). *Typologies and taxonomies: an introduction to classification techniques* (Vol. 102), Sage.

Bartlett, L., & Vavrus, F. (2017). Comparative Case Studies: An Innovative Approach. *Nordic Journal of Comparative and International Education (NJCIE), 1*(1), 5–17.

Benson-Rea, M., Brodie, R. J., & Sima, H. (2013). The plurality of co-existing business models: Investigating the complexity of value drivers. *Industrial Marketing Management, 42*(5), 717–729.

Betz, F. (2002). Strategic business models. *Engineering Management Journal, 14*(1), 21–28.

Bienstock, C., Gillenson, M., & Sanders, T. (2002). The complete taxonomy of web business models. *Quarterly Journal of Electronic Commerce, 3*, 173–186.

Bierman, A. S., Pugh, M. J. V., Dhalla, I., Amuan, M., Fincke, B. G., Rosen, A., et al. (2007). Sex differences in inappropriate prescribing among elderly veterans. *The American Journal of Geriatric Pharmacotherapy, 5*(2), 147–161.

Brady, M. P., & Saranga, H. (2013). Innovative business models in healthcare: a comparison between India and Ireland. *Strategic Change, 22*(5–6), 339–353.

Bronowski, J., & Long, W. (1951). Statistical methods in anthropology. *Nature, 168*(4279), 794.

Calás, M. B., Smircich, L., & Bourne, K. A. (2009). Extending the Boundaries: Reframing "Entrepreneurship as Social Change" Through Feminist Perspectives. *Academy of Management Review, 34*(3), 552–569.

Campbell, S. (2010). Comparative Case Study. In A. J. Mills, G. Durepos, & E. Wiebe (Eds.), *Encyclopedia of Case Study Research*. Thousand Oaks, California: SAGE Publications, Inc. 175–176.

Casadesus-Masanell, R., & Ricart, J. E. (2010). From strategy to business models and onto tactics. *Long Range Planning, 43*(2–3), 195–215.

Casadesus-Masanell, R., & Zhu, F. (2013). Business model innovation and competitive imitation: The case of sponsor-based business models. *Strategic Management Journal, 34*(4), 464–482.

Chilet-Rosell, E. (2014). Gender bias in clinical research, pharmaceutical marketing, and the prescription of drugs. *Global Health Action, 7*(1), 25484.

CIHR, N. (2010). SSHRC (Canadian Institutes of Health Research, Natural Sciences and Engineering Research Council of Canada and Social Sciences and Humanities Research Council of Canada), 2010 Tri-Council Policy Statement: Ethical Conduct For Research Involving Humans. *Tri-Council policy statement: Ethical conduct for research involving humans*.

Clayton, J. A., & Collins, F. S. (2014). NIH to balance sex in cell and animal studies. *Nature, 509* (7500), 282–283.

Coen, S., & Banister, E. (2012). What a difference sex and gender make: a gender, sex and health research casebook. *Ottawa, Canada: Canadian Institutes of Health Research*.

DaSilva, C. M., & Trkman, P. (2014). Business model: What it is and what it is not. *Long Range Planning, 47*(6), 379–389.

Day, S., Mason, R., Lagosky, S., & Rochon, P. A. (2016). Integrating and evaluating sex and gender in health research. *Health Research Policy and Systems, 14*(1), 75.

Dhar, N., Chaturvedi, S., & Nandan, D. (2011). Spiritual health scale 2011: defining and measuring 4 dimension of health. *Indian journal of community medicine : official publication of Indian Association of Preventive & Social Medicine, 36*(4), 275–282.

Doyal, L. (2001). Sex, gender, and health: the need for a new approach. *BMJ: British Medical Journal, 323*(7320), 1061.

Dubosson-Torbay, M., Osterwalder, A., & Pigneur, Y. (2002). E-business model design, classification, and measurements. *Thunderbird International Business Review, 44*(1), 5–23.

EC (2003a). "*She figures*": *women and science statistics and indicators*: Office for Official Publications of the European Communities.

EC (2003b). Vademecum: Gender mainstreaming in the Sixth Framework Programme—Reference guide for scientific officers and project officers. Directorate-General for Research Brussels.

EC (2013). Gendered innovations: How gender analysis contributes to research. European Commission. Eur 25848.

Ekbia, H. R. (2009). Digital artifacts as quasi-objects: Qualification, mediation, and materiality. *Journal of the Association for Information Science and Technology*, 60(12), 2554–2566.

Eland-de Kok, P., van Os-Medendorp, H., Vergouwe-Meijer, A., Bruijnzeel-Koomen, C., & Ros, W. (2011). A systematic review of the effects of e-health on chronically ill patients. *Journal of Clinical Nursing*, 20(21-22), 2997–3010.

Eysenbach, G., & Jadad, A. R. (2001). Evidence-based patient choice and consumer health informatics in the Internet age. *Journal of Medical Internet Research*, 3(2).

Fausto-Sterling, A. (2012). *Sex/gender: Biology in a social world*: Routledge.

Fishman, J. R., Wick, J. G., & Koenig, B. A. (1999). The use of "sex" and "gender" to define and characterize meaningful differences between men and women. *Agenda for research on women's health for the 21st century*. Washington DC: Department of Health and Human Services.

Fjeldstad, Ø. D., & Snow, C. C. (2018). Business models and organization design. *Long Range Planning*, 51(1), 32–39.

Fox, S. (2018). Domesticating artificial intelligence: Expanding human self-expression through applications of artificial intelligence in prosumption. *Journal of Consumer Culture*, 18(1), 169–183.

Fredriksson, J. J., Mazzocato, P., Muhammed, R., & Savage, C. (2017). Business model framework applications in health care: A systematic review. *Health Services Management Research*, 30(4), 219–226.

Gahagan, J., Gray, K., & Whynacht, A. (2015). Sex and gender matter in health research: addressing health inequities in health research reporting. *International Journal for Equity in Health*, 14(1), 12.

GenSET (2010). Recommendations for action on the gender dimension in science. Portia London.

George, G., & Bock, A. J. (2011). The business model in practice and its implications for entrepreneurship research. *Entrepreneurship Theory and Practice*, 35(1), 83–111.

Gordijn, J., Osterwalder, A., & Pigneur, Y. (2005). Comparing two business model ontologies for designing e-business models and value constellations. *BLED 2005 Proceedings*, 15.

Greaves, L. (2011). Why put gender and sex into health research. *Designing and conducting Gender, Sex & Health Research*, 3–14.

Gulati, R., & Soni, T. (2015). Digitization: A strategic key to business. *Journal of Advances in Business Management*, 1(2), 60–67.

Hambrick, D. C. (1984). Taxonomic approaches to studying strategy: Some conceptual and methodological issues. *Journal of Management*, 10(1), 27–41.

Haverman, L., Engelen, V., van Rossum, M. A., Heymans, H. S., & Grootenhuis, M. A. (2011). Monitoring health-related quality of life in paediatric practice: development of an innovative web-based application. *BMC Pediatrics*, 11(1), 3.

Haverman, L., Grootenhuis, M. A., van den Berg, J. M., van Veenendaal, M., Dolman, K. M., Swart, J. F., et al. (2012). Predictors of health-related quality of life in children and adolescents with juvenile idiopathic arthritis: Results from a web-based survey. *Arthritis Care & Research*, 64(5), 694–703.

Hawkins, R. (2002). The phantom of the marketplace: searching for new e-commerce business models. *Euro CPR*, 24–26.

Hedman, J., & Kalling, T. (2003). The business model concept: theoretical underpinnings and empirical illustrations. *European Journal of Information Systems*, 12(1), 49–59.

Henry, C., Foss, L., & Ahl, H. (2016). Gender and entrepreneurship research: A review of methodological approaches. *International Small Business Journal*, 34(3), 217–241.

Hoc, J.-M. (2000). From human–machine interaction to human–machine cooperation. *Ergonomics*, *43*(7), 833–843.

Howcroft, D. (2001). After the goldrush: deconstructing the myths of the dot. com market. *Journal of Information Technology*, *16*(4), 195–204.

Jeffrey, C. (1982). Kingdoms, codes and classification. *Kew Bulletin*, 403–416.

Jennings, J., & Brush, C. (2013). Research on women entrepreneurs: Challenges to (and from) the broader entrepreneurship literature? *The Academy of Management Annals*, *7*(1), 663–715.

Judd, T. (2018). The rise and fall (?) of the digital natives. *Australasian Journal of Educational Technology*, *34*(5).

Kauffman, R. J., & Wang, B. (2008). Tuning into the digital channel: evaluating business model characteristics for Internet firm survival. *Information Technology and Management*, *9*(3), 215–232.

Klagge, B., & Meister, T. (2018). Energy cooperatives in Germany–an example of successful alternative economies? *Local Environment*, *23*(7), 697–716.

Kondratieff, N. D., & Stolper, W. F. (1935). The Long Waves in Economic Life. *The Review of Economics and Statistics*, *17*(6), 105–115.

Krieger, N. (2003). Genders, sexes, and health: what are the connections—and why does it matter? *International Journal of Epidemiology*, *32*(4), 652–657.

Lambert, S. C. (2015). The importance of classification to business model research. *Journal of Business Models*, *3*(1), 49.

Lee, C., Tsenkova, V. K., Boylan, J. M., & Ryff, C. D. (2018). Gender differences in the pathways from childhood disadvantage to metabolic syndrome in adulthood: An examination of health lifestyles. *SSM-population Health*, *4*, 216–224.

Lyytinen, K., Yoo, Y., & Boland Jr, R. J. (2016). Digital product innovation within four classes of innovation networks. *Information Systems Journal*, *26*(1), 47–75.

Mair, F. S., May, C., O'Donnell, C., Finch, T., Sullivan, F., & Murray, E. (2012). Factors that promote or inhibit the implementation of e-health systems: an explanatory systematic review. *Bulletin of the World Health Organization*, *90*(5), 357–364.

Marquez, J. J., Riley, V., & Schutte, P. C. (2018). Chapter 10 - Human automation interaction A2 - Sgobba, Tommaso. In B. Kanki, J.-F. Clervoy, & G. M. Sandal (Eds.), *Space Safety and Human Performance* (pp. 429–467): Butterworth-Heinemann.

McKelvey, B. (1982). *Organizational systematics--taxonomy, evolution, classification*: Univ of California Press.

Min, H., Caltagirone, J., & Serpico, A. (2008). Life after a dot-com bubble. *International Journal of Information Technology and Management*, *7*(1), 21–35.

Muntean, S. C., & Ozkazanc-Pan, B. (2015). A gender integrative conceptualization of entrepreneurship. *New England Journal of Entrepreneurship*, *18*(1), 27–40.

Nambisan, S. (2017). Digital entrepreneurship: Toward a digital technology perspective of entrepreneurship. *Entrepreneurship Theory and Practice*, *41*(6), 1029–1055.

Nambisan, S., Lyytinen, K., Majchrzak, A., & Song, M. (2017). Digital innovation management: Reinventing innovation management research in a digital world. *Mis Quarterly*, *41*(1), 223–238.

Nefiodow, L., & Nefiodow, S. (2014). *The Sixth Kondratieff: The New Long Wave in the Global Economy*.

OECD (2017). *Health at a Glance 2017*.

Osterwalder, A. (2004). The business model ontology: A proposition in a design science approach.

Osterwalder, A., Pigneur, Y., & Tucci, C. L. (2005). Clarifying business models: Origins, present, and future of the concept. *Communications of the Association for Information Systems*, *16*(1), 1.

Pateli, A. G., & Giaglis, G. M. (2004). A research framework for analysing eBusiness models. *European Journal of Information Systems*, *13*(4), 302–314.

Paton, C., Hansen, M., Fernandez-Luque, L., & Lau, A. Y. (2012). Self-tracking, social media and personal health records for patient empowered self-care. *Yearbook of Medical Informatics*, 21(01), 16–24.

Porter, D. (1999). The history of public health: current themes and approaches. *Hygiea Internationalis*, 1(1), 9–21.

Prensky, M. (2001). Digital natives, digital immigrants part 1. *On the Horizon*, 9(5), 1–6.

Qureshi, S., & Keen, P. (2005). Activating knowledge through electronic collaboration: Vanquishing the knowledge paradox. *IEEE Transactions on Professional Communication*, 48(1), 40–54.

Rajala, R., & Westerlund, M. (2007). Business models–a new perspective on firms' assets and capabilities: observations from the Finnish software industry. *The International Journal of Entrepreneurship and Innovation*, 8(2), 115–125.

Ricciardi, F., Zardini, A., & Rossignoli, C. (2016). Organizational dynamism and adaptive business model innovation: The triple paradox configuration. *Journal of Business Research*, 69(11), 5487–5493.

Ritter, T., & Lettl, C. (2017). The wider implications of business-model research. *Long Range Planning*, 51(1), 1–8.

Schiebinger, L. (2014b). Scientific research must take gender into account. *Nature*, 507(7490), 9.

Schiebinger, L., & Klinge, I. (2015). Gendered innovation in health and medicine. *GENDER–Zeitschrift für Geschlecht, Kultur und Gesellschaft*, 7(2), 29–50.

Schiebinger, L., Klinge, I., Paik, H. Y., Sánchez de Madariaga, I., Schraudner, M., & Stefanick, M. (2011–2017). *Gendered Innovations in Science, Health & Medicine, Engineering, and Environment (genderedinnovations.stanford.edu)*.

Schiebinger, L., & Schraudner, M. (2011). Interdisciplinary Approaches to Achieving Gendered Innovations in Science, Medicine, and Engineering1. *Interdisciplinary Science Reviews*, 36(2), 154–167.

Schumpeter, J. A. (1939). *Business cycles* (Vol. 1): McGraw-Hill New York.

Schumpeter, J. A. (1961). *The theory of economic developmet*: Oxford University Press.

Seong Leem, C., Sik Suh, H., & Seong Kim, D. (2004). A classification of mobile business models and its applications. *Industrial Management & Data Systems*, 104(1), 78–87.

Shepherd, D. A., & Patzelt, H. (2015). The "heart" of entrepreneurship: The impact of entrepreneurial action on health and health on entrepreneurial action. *Journal of Business Venturing Insights*, 4, 22–29.

Simpson, G. G. (1961). *Principles of animal taxonomy*. Columbia: Colmbia University Press.

Simpson, G. G. (1964). Numerical taxonomy and biological classification. American Association for the Advancement of Science.

Smith, E. E., & Medin, D. L. (1981). *Categories and concepts* (Vol. 9): Harvard University Press Cambridge, MA.

Sneath, P. H., & Sokal, R. R. (1973). *Numerical taxonomy. The principles and practice of numerical classification*.

Snyder, C. F., Aaronson, N. K., Choucair, A. K., Elliott, T. E., Greenhalgh, J., Halyard, M. Y., et al. (2012). Implementing patient-reported outcomes assessment in clinical practice: a review of the options and considerations. *Quality of Life Research*, 21(8), 1305–1314.

SoKAL, R. R., & Sneath, P. H. (1963). Principles of numerical taxonomy. 359 pp. *San Francisco*.

Spieth, P., Schneckenberg, D., & Ricart, J. E. (2014). Business model innovation–state of the art and future challenges for the field. *R&D Management*, 44(3), 237–247.

Stephan, U. (2018). Entrepreneurs' Mental Health and Well-Being: A Review and Research Agenda. *Academy of Management Perspectives*, 32(3), 290–322.

Swan, M. (2012). Health 2050: The realization of personalized medicine through crowdsourcing, the quantified self, and the participatory biocitizen. *Journal of Personalized Medicine, 2*(3), 93–118.

Tapscott, D. (2000). *Digital capital: Harnessing the power of business webs*: Harvard Business School Press.

Teece, D. J. (2010). Business models, business strategy and innovation. *Long Range Planning, 43*(2–3), 172–194.

Timmers, P. (1998). Business models for electronic markets. *Electronic Markets, 8*(2), 3–8.

Tiwana, A., Konsynski, B., & Bush, A. A. (2010). Research commentary—Platform evolution: Coevolution of platform architecture, governance, and environmental dynamics. *Information Systems Research, 21*(4), 675–687.

UNESCO (2007). Science, technology, and gender: An international report. UNESCO Paris.

Valderas, J., Kotzeva, A., Espallargues, M., Guyatt, G., Ferrans, C., Halyard, M. Y., et al. (2008). The impact of measuring patient-reported outcomes in clinical practice: a systematic review of the literature. *Quality of Life Research, 17*(2), 179–193.

Valliere, D., & Peterson, R. (2004). Inflating the bubble: examining dot-com investor behaviour. *Venture Capital, 6*(1), 1–22.

von Briel, F., Davidsson, P., & Recker, J. (2018). Digital technologies as external enablers of new venture creation in the IT hardware sector. *Entrepreneurship Theory and Practice, 42*(1), 47–69.

Warriner, C. K. (1984). *Organizations and their environments: Essays in the sociology of organizations* (Vol. 3): Greenwich, Conn.: Jai Press.

Weill, P., & Vitale, M. (2002). What IT infrastructure capabilities are needed to implement e-business models? *Mis Quarterly, 1*(1), 17.

WHO, W. H. O. (1946). Consitution of the World Health Organization. *Off. Rec. World Health Organisation, 2*(100).

Wicks, P., Stamford, J., Grootenhuis, M. A., Haverman, L., & Ahmed, S. (2014). Innovations in e-health. *Quality of Life Research, 23*(1), 195–203.

Wicks, P., Vaughan, T. E., Massagli, M. P., & Heywood, J. (2011). Accelerated clinical discovery using self-reported patient data collected online and a patient-matching algorithm. *Nature Biotechnology, 29*(5), 411.

Wiener, N. (1948). Cybernetics. *Scientific American, 179*(5), 14–19.

Winterhalter, S., Zeschky, M. B., Neumann, L., & Gassmann, O. (2017). Business Models for Frugal Innovation in Emerging Markets: The Case of the Medical Device and Laboratory Equipment Industry. *Technovation, 66*, 3–13.

Wirmanm, H. (2014). Games for/with strangers - Captive orangutan (pongo pygmaeus) touch screen play. *Aentennae: The Journal of Nature in Visual Culture, 30*, 04–11.

Yamamoto, Y. (2018). An Organizational Form for the Development of Renewable Energy. In *Feed-in Tariffs and the Economics of Renewable Energy* (pp. 151–159): Springer.

Zott, C., Amit, R., & Massa, L. (2011). The business model: recent developments and future research. *Journal of Management, 37*(4), 1019–1042.

List of Figures

Figure 2.1	Overview of the experiments (Source: Own illustration)	—— 14
Figure 2.2	Results of the second group (Source: Own illustration)	—— 16
Figure 2.3	Results of the third group (Source: Own illustration)	—— 17
Figure 2.4	Results of the fourth group (Source: Own illustration)	—— 18
Figure 2.5	General overview (Source: Own illustration) —— 19	
Figure 2.6	Results according to color (Source: Own illustration) —— 20	
Figure 3.1	Results in a non-competitive setting, men (Source: Own illustration) —— 36	
Figure 3.2	Results in a non-competitive setting, women (Source: Own illustration) —— 36	
Figure 3.3	Results in a competitive setting, men (Source: Own illustration) —— 37	
Figure 3.4	Results in a competitive setting, women (Source: Own illustration) —— 38	
Figure 4.1	Technology Acceptance Model (Source: Own illustration based on Davis 1989) —— 45	
Figure 4.2	Chair used in the experiment with functions; 1 = seat height; 2 = backrest; 3 = seat depth; 4 = lumbar support; 5 = armrest height; 6 = armrest depth; 7 = seat width; 8 = backrest pressure (Source: Own illustration) —— 49	
Figure 4.3	Frequently used chairs (Source: [Chair A59], n.d.; [LÅNGFJÄLL], n.d.; [Ikea Volmar + Armlehne], n.d.) —— 50	
Figure 6.1	Network diagram of the measures that were tested in the experiments (Source: Own illustration) —— 79	
Figure 6.2	Characteristics of the instructions in connection with the MRT-results (Source: Own illustration) —— 81	
Figure 7.1	Heuristic scheme to support gender-aware business model classification (Source: Own illustration) —— 96	

Index

Alternate uses task 30
Assisting materials 3, 63, 64, 65, 66, 67, 69

Buoyant device 3, 63, 64

Classification of business models 94, 95
Cognitive processes 2, 8, 10, 46, 60
Color fixation 19
Competitive behaviour 3
Competitive environments 5, 29, 32, 37, 39, 40
Crafting 4, 13, 15, 20, 24, 25
Creativity 2, 4, 7, 8, 9–10, 12, 14, 21, 22, 23, 24, 26, 29, 30–31, 33, 34, 38, 39, 40, 41, 73, 90

Design fixation 10, 11
Desk chair 77, 83
Digital Business Models 4, 88, 95, 97
Digital shift 91, 92
Digital technology 88, 91
Digital transformation 4, 88, 89–93, 95
Digitalization 91, 92, 93, 95
Digitization 90, 91, 92
Divergent thinking 29, 30–31, 34, 38
Dummy object 3, 63, 64

Educational and occupational orientation 58
E-health 4, 88, 92, 93–97, 98
Entrepreneurship 1, 4, 29, 88, 89–90, 92, 93, 95, 97
Examples 2, 7, 8, 9, 12, 17, 19, 21, 22, 23, 25, 26, 46, 59, 92
Experiment 3, 4, 9, 11, 12, 13, 14, 15, 16, 17, 18, 19, 21, 22, 23, 24, 25, 26, 29, 34, 35, 36, 37, 38, 39, 40, 43, 44, 45, 46, 47, 48, 49, 50, 52, 53, 54, 58, 61, 62, 63, 64–65, 66, 67, 69, 73, 75, 76, 78, 80, 82, 83
Experimental research project 73
Experimental setup 15, 48, 58, 77

Fearfulness 46, 47
Female underrepresentation 57
Fixation effect 2, 8, 10–12, 18, 19, 21, 22, 23, 26
Forming clay 3, 62, 63, 64
Functional fixation 10, 23, 25

Gender 1–5, 8, 10, 13, 15, 16, 17, 18, 20, 21, 22, 23, 25, 26, 29, 30–33, 34, 35, 37, 38, 39, 41, 43, 44–47, 68, 73, 74, 75, 76, 79, 80, 81, 83, 87, 88, 89–90, 93–97, 98
Gender analysis 87, 89, 90, 93, 95, 97
Gender awareness 2, 4, 88, 89, 90, 93–97, 98
Gender differences 1–5, 8, 10, 22, 23, 25, 29, 31, 32, 33, 43, 44, 45, 47, 48, 50, 54, 65, 66–68, 73, 74, 76, 83, 89
Gender differences in competitive behaviour 2, 29, 31–33, 34, 37, 39, 40
Gender differences in the intuitive usage of products 3, 46–47
Gender gap 3, 5, 29, 32, 33, 54, 55, 74, 81
Gender imbalance 57, 58–59
Gendered innovation potential 4, 88

Health care sector 90, 92, 93, 95
Heuristic scheme 4, 88, 93–97, 98

Idea creation 9, 12, 33, 35, 39, 93
Idea generation 2, 7, 9, 11, 26
Ideal choice of assisting materials 64, 65
Information technology 88
Innovation 1–5, 7, 9, 30, 43, 69, 73, 87, 88, 89–90, 92, 93, 94, 95, 98
Innovative strength 73
Instruction design 76
Instructions 3, 4, 13, 14, 15, 21, 23, 26, 47, 48, 54, 61, 63, 65, 69, 75, 76, 77, 78, 79, 80, 81, 82, 83
Intuition 46

Likert scale 51, 77

Mental rotation test (MRT) 60, 75, 77, 80, 81
Methodological problem-solving 3, 59–61, 62, 66, 67–68
Modeling clay 13, 14, 15, 16, 18, 22, 23, 25, 26

New business models 87

Office chair functions 3, 53
Originality 10, 11, 30, 35

People with different educational levels 1
Prior experience with products 43, 47, 48
Problem-solving 2, 4, 7, 8, 10, 11, 23, 58, 59–61, 62, 64, 65, 66, 67–68, 69, 70, 74, 82, 83
Problem-solving skills 3, 4, 74
Public health 87, 88

Quasi-experiment 3, 24, 40, 43, 47, 48, 54, 55, 68, 69, 76, 77, 83

Risk-taking 46, 47
Rod and Frame test 60

Spatial perception test 60
Spatial reasoning abilities 3, 4, 58, 59, 61, 62, 64, 65, 66–68, 69, 70
Spatial relations 60, 66, 74
Spatial thinking 4, 44, 60, 74, 75, 76, 77, 80, 81, 82

Spatial visualisation abilities 59, 60, 61, 74
Spoken example 15
STEM disciplines 3, 69, 73
Stereotypical gender roles 23, 57

Task solving 66, 69
Technical tasks 4, 73, 76, 80, 83
Technical understanding 76, 81
Technology Acceptance Model 45, 50, 51–52, 53
Technology management 2, 5, 43, 73
Testing method 60, 74
Theory of Planned Behaviour 45
Theory of Reasoned Action 45

Valid reflective self-assessment 64, 65, 67
Value creation 4, 87, 88, 90, 91–93, 98
Verbal abilities 66, 75
Visual example 12, 16, 21

List of Tables

Table 2.1	Overview of the test subjects (Source: Own illustration based on Agogué and Cassotti, 2013, p. 6) —— 13
Table 2.2	Distribution of the genders (Source: own illustration based on Agogué and Cassotti, 2013, p. 6) —— 13
Table 3.1	Excerpt of our evaluation sheet (Source: Own illustration) —— 35
Table 4.1	Functions found by the participants (Source: Own illustration) —— 51
Table 4.2	Components of the Technology Acceptance Model and the participants' answers (Source: Own illustration) —— 51
Table 4.3	Prior experience of the participants with the different chair types (Source: Own illustration) —— 52
Table 5.1	Results from the experiment (Source: Own illustration) —— 65
Table 7.1	Contrasting digitization, digitalization, digital shift and digital Transformations (Source: Own illustration) —— 92
Table 7.2	E-Health examples of digital modus operandi (Source: Own illustration) —— 93